CHICKEN
BEST EVER POULTRY RECIPES

CHICKEN
BEST EVER POULTRY RECIPES

bay books

MEAL PLANNER

Use the following table to plan your best-ever Chicken meal. The recipes have been grouped into appropriate classifications and the portion size of each recipe is clearly shown. Plan your meal and then turn to the appropriate page to find your clear and concise recipe together with a large-format picture of the finished dish.

SNACKS & STARTERS

RECIPE	PAGE	PORTIONS
Crispy chicken wings	12	6
Spicy buffalo wings with ranch dressing	15	4
Chicken and lime hummus tortillas	19	4
Chicken fajitas	23	4
Chicken quesadillas	24	4
Chicken and tzatziki wrap	32	4
Chicken tikka	35	25 skewers
Satay chicken	39	4
Chinese mushroom and chicken soup	43	4
Smoked chicken caesar salad	47	4
Thai-style chicken and baby corn soup	48	4
Tom kha gai	51	4
Chilli chicken and cashew salad	52	4
Indian marinated chicken salad	55	4
Chicken and vegetable soup	64	4–6
Chicken and pasta salad	67	4
Chunky chicken and vegetable soup	68	4
Spicy chicken broth with coriander and pasta	72	4
Bacon-wrapped chicken	78	6
Blackened chicken with crispy tortilla	149	4
Miso yakitori chicken	153	4
Wild rice and roast chicken with Asian dressing	154	8
Avgolemono with chicken	169	4

BAKES & BARBECUES

RECIPE	PAGE	PORTIONS
Chicken burger with tarragon mayonnaise	16	6
Chicken burger with tangy garlic mayonnaise	20	4
Low-fat chicken pies	27	4
Yakitori	31	25 skewers
Chicken and corn pies	36	6
New potato, chicken and spinach frittata	44	4–6
Five spice roast chicken	81	4
Honey mustard chicken	82	4–6
Lebanese chicken	85	4–6
Margarita chicken	86	4
Mirin and sake chicken	89	4
Sage and ricotta stuffed chicken	90	4
Thai spiced chicken with potato rosti	94	6
Moroccan chicken	101	4
Chicken and bacon gougere	121	6
Chicken and leek pie	122	4
Chicken ballotine	125	8
Family chicken pie	133	6
Orange roasted chickens	141	8
Chargrilled chicken with spinach and raspberries	150	4
Chicken breasts with mustard cream sauce	157	4
Mexican chicken bake	158	4
Spanish saffron chicken and rice	161	4

CASSEROLES

CURRIES

STIR-FRIES & PAN-FRIES

CHICKEN WITH PASTA

CHICKEN BASICS

Chicken is a versatile food, which lends itself to many different types of recipes, methods of cooking and styles of cuisine. This comprehensive guide to cooking with chicken, including information on the different types of chicken cuts available, purchasing, storing and cooking chicken, will help you to prepare successful and delicious chicken dishes.

Here is advice on how to get the best out of chicken: what to look for when buying and how to store and prepare your purchase prior to cooking. There's also information on how to make stock and gravy, jointing and boning chickens, and the art of carving.

PURCHASING

There is a wide variety of chickens and chicken cuts available.

Young tender birds are good for grilling, barbecuing, frying and roasting. Chickens over 1.8 kg (3 lb 10 oz) are usually only suitable for poaching, braising or boiling because of their tough flesh (although they have excellent flavour) and are labelled as boiling fowls. Unless you specifically want a very large bird, buy two smaller chickens rather than one large.

Whole baby chickens (or poussins) weigh about 500 g (1 lb) and serve one person. Chickens under 1.5 kg (3 lb) serve two to four people. Whole roasting chickens weighing over 1.5 kg (3 lb) should serve at least four people.

Chickens are sold cleaned, with their innards removed. The neck is usually tucked inside. Sometimes the giblets are in the cavity, contained in a plastic bag, so be sure to remove them before cooking or freezing.

Whole birds are marketed by weight, and the weight becomes the number of the chicken. For example, a No. 10 chicken weighs 1 kg (2 lb); a No. 16 chicken weighs 1.6 kg (3$^1/_4$ lb), and so on.

When buying chicken by weight, check whether the giblets are with the bird. Giblets weigh approximately 175 g (6 oz) and will affect the number of portions you get after cooking.

FRESH CHICKEN

Fresh chicken has better flavour and texture than frozen. Look for skin that is light pink and moist, rather than wet, with no dry spots. It should be unbroken and free from blemishes and bruises. The breast should be plump and well rounded; on a young bird the point of the breastbone will be flexible.

At speciality poultry shops you can buy free-range, grain-fed, and corn-fed chickens (with yellow skin and flesh).

Chicken can also be purchased cooked. Hot take-away barbecued, roasted or chargrilled

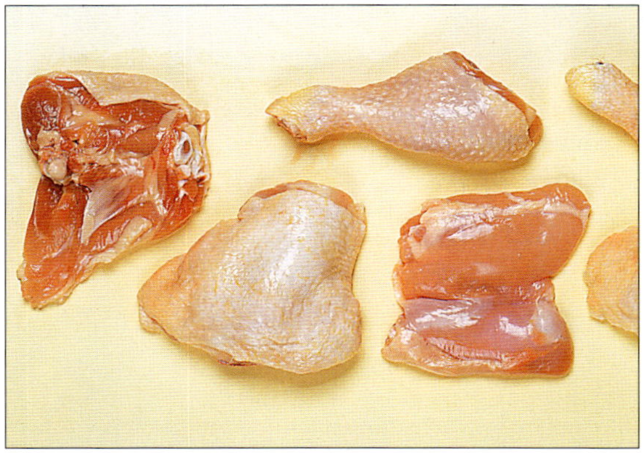

Chicken leg cuts, clockwise from left: thigh (underside view), drumstick, maryland, thigh fillet, thigh.

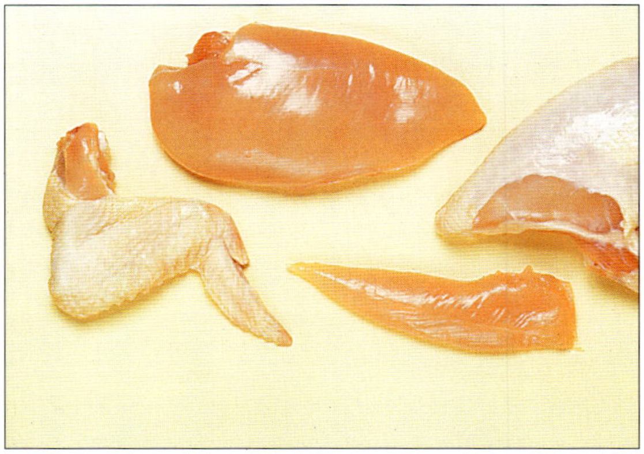

Chicken breast and wing cuts, clockwise from left: wing, single breast fillet, whole breast with bone, tenderloin.

chicken has become a mainstay of busy people as the basis for quick and satisfying meals. Cooked, smoked chicken is available whole, chilled, from delicatessens and supermarkets.

Chicken cuts include: double or single breasts on the bone, with skin or without; breast fillets; tenderloins (the part just behind the breast); marylands (the whole thigh and leg); thigh cutlets; thigh fillets; wings; and drumsticks (the bottom part of the leg). Buy the appropriate cut of chicken for the cooking process you will be using. Chicken is very versatile, and it is not always necessary to buy the most expensive fillets to produce an excellent result. Here is a guide to the cuts used in this book:

FOR ROASTING – whole roasting chickens, baby chickens, whole breasts, wings, marylands, drumsticks, thighs.

FOR GRILLING – chicken halves and quarters, wings, drumsticks, marylands, thigh cutlets.

FOR BARBECUING – chicken halves, whole breasts, wings, drumsticks, marylands, thigh cutlets, tenderloins.

FOR STIR-FRYING – breast fillets, thigh fillets, tenderloins, livers.

FOR PAN-FRYING – marylands, breast fillets, tenderloins, livers.

FOR DEEP-FRYING – drumsticks, wings, thighs, chicken pieces.

FOR CASSEROLES/BRAISING – whole chickens, chicken pieces, thighs, thigh cutlets, drumsticks, wings.

FOR POACHING – whole chickens, whole breasts, breast fillets, thighs, drumsticks.

FOR STOCK – bones, necks, giblets, boiling fowls.

STORING FRESH CHICKEN

Chicken must be transported home as quickly as possible. Do not leave it sitting in the sun in the car or car boot. The internal temperature of a car left closed in full sun spells disaster to all chicken products. The longer that food spends between 5°C (41°F) and 60°C (140°F), the greater the likelihood of rapid growth of harmful bacteria that may result in food poisoning.

Keep chicken away from any strong-smelling items such as cleaning agents and petrol that you may have stored in your vehicle; chicken will absorb the smells.

Make it your policy to purchase chicken or meat as the last item on your round of shopping. In hot weather, use an insulated chiller bag to keep it cold.

It is particularly important to store poultry carefully to avoid contamination by salmonella bacteria, which can cause food poisoning. Always wash hands, chopping boards, knives and cooking implements in very hot soapy water after handling raw chicken. Always keep cooked and raw chicken separate.

Before storing uncooked whole chicken, discard the tight plastic wrappings and pour off any juices. Remove the neck and giblets from whole birds (sometimes these are in a plastic bag inside the cavity). Giblets should be cooked immediately or stored separately. Use the neck and giblets for stock; chop the liver to flavour a sauce, gravy or stuffing.

Loosely wrap the chicken in plastic wrap or place in a plastic bag, place the package on a plate and refrigerate on the bottom shelf of the refrigerator. Never place uncooked chicken where the juices could drip on or otherwise come into contact with other foodstuffs. A fresh, cleaned and wrapped chicken can be stored in the refrigerator for up to two days.

To store fresh chicken pieces or cuts, remove them from trays or other packaging, pour off any juices and loosely wrap in plastic wrap or place in a plastic bag. Refrigerate for up to two days.

STORING COOKED CHICKEN

Chicken should stand no more than an hour at room temperature after cooking. If keeping longer than this, store it loosely wrapped in the refrigerator and use within three days. The chicken does not have to be cold when it goes into the refrigerator. If the chicken has a sauce or stuffing, it should be eaten within 24 hours of cooking. Stuffing and gravy should be stored separately.

FROZEN CHICKEN

Make sure that any frozen chicken you buy from the supermarket freezer is solid and completely enclosed in its packaging. Do not purchase any that appear semi-soft and that are sitting in their own juices, as this indicates that they have been in the display cabinet for longer than is ideal.

Uncooked, home-frozen chicken (without giblets) will keep for up to nine months in good condition. Remove the giblets before freezing as they will begin to deteriorate after eight weeks.

If a package has partly defrosted it must never be refrozen; defrost fully in the refrigerator and cook promptly. Stuffed birds should never be frozen, as the filling will not freeze enough to prevent the development of harmful bacteria.

Freezing fresh chicken: Have the freezer temperature at minus 15°C (5°F) or lower. Use heavy-gauge polythene bags and good-quality plastic wrap to package fresh chicken.

Label each package with the details of the contents, the date it was packaged and stored and either the number of people it will feed or the unfrozen weight. Use a waterproof pen or wax crayon.

It is important to expel as much air as possible from the packaging; oxygen left behind will speed up the process of oxidisation of any fat, resulting in an unpleasant taste after prolonged storing. If you do a lot of freezing, it may be worthwhile investing in a vacuum freezer pump to efficiently expel air.

Secure freezer bags by twisting the tips and closing them with masking tape; this is preferable to metal twist ties. Or, clip a metal band in place with a clipping device.

Wrap the chicken in heavy-gauge freezer bags or good-quality plastic wrap, then in aluminium foil, expelling air. Label and freeze. Remove the giblets and neck and pack separately to be used for stock.

Commercially frozen whole chickens without giblets can be kept for nine months, or according to instructions on the package. If the chicken is still frozen hard, place it in the freezer in its original wrappings as soon as possible after purchase. If it has started to thaw, place it in the refrigerator to completely thaw out, cook it promptly, cool quickly, wrap and refreeze it. Never refreeze thawed, uncooked poultry.

To freeze uncooked chicken, wrap in heavy-duty plastic bags.

To freeze cooked dishes, place into plastic or aluminium containers.

Whole chickens can be trussed before freezing to be ready for cooking when thawed. To provide convenient serving portions and to save space, cut the chicken into four, eight or 10 pieces before freezing. Wrap each joint individually in plastic wrap, expelling air, and then combine in a larger package in a sturdy plastic bag. Seal, tape the end of the bag to the package, label and freeze.

Chicken cuts such as legs or wings can be frozen either as individual portions or in amounts to serve several people. Wrap portions individually in plastic wrap, expelling air, then place them in a larger plastic bag. Seal and tape the end of the bag to the package. Label and freeze.

Boneless cuts can be cut into strips or cubes before freezing. Weigh out meal-sized portions and place in heavy-duty plastic bags. Fill right to the corners with chicken. Flatten the package (so it will defrost quickly) and expel the air. Seal, label and freeze.

The same procedure applies to chicken mince.

FREEZING COOKED CHICKEN

Cooked whole chickens or chicken pieces can be frozen with or without bones, for up to two weeks. After this time it will tend to dry out.

Moist chicken dishes such as stews, casseroles, curries and soups are all suitable for freezing. Quickly reduce the temperature of the cooked item by placing it in the refrigerator or by plunging the base of the dish into cold water, then cool completely in the refrigerator. Line cake tins or other suitable containers with a heavy-duty plastic bag. Spoon portions into the bag and expel the air. Seal, label and freeze. Place the bag in its tin into the freezer. When frozen, remove from the tin, reseal to remove as much air as possible and return, labelled, to the freezer.

Or, simply spoon the meat directly into plastic or aluminium containers and seal, label and freeze. As a general rule, freeze cooked chicken for a maximum of two months.

DEFROSTING

Frozen chicken must be completely thawed in the refrigerator before cooking; allow two to three hours per 500 g (1 lb). A frozen chicken should be cooked within 12 hours of thawing. Do not thaw chicken at room temperature. Never keep perishable food at room temperature for longer than two hours, particularly on hot days. This includes time to prepare, serve and eat.

Microwave defrosting is not recommended for whole frozen chickens because of uneven thawing. However, smaller packages of cuts or

Trim any pockets of fat from the chicken cavity before cooking.

To marinate, pour the liquid over the chicken in a non-metallic dish.

pre-cooked meals can be successfully thawed in the microwave using the defrost setting. Always remove chicken from wrapping before defrosting. Stir casseroles occasionally to distribute heat evenly. Separate joints or pieces as they soften.

PREPARATION

Before cooking a chicken, remove the neck, giblets and fat pockets from the cavity. Discard the fat and use the neck and giblets for stock. Remove any excess fat and sinew from chicken pieces.

Raw poultry should be wiped with a damp cloth, rather than washed, before cooking. Wipe frozen chickens with paper towels to absorb excess moisture.

Use a cook's knife for jointing uncooked chickens. Poultry shears are excellent for dividing whole chickens into serving portions, expecially for splitting the breastbone, cutting the backbone and rib bones and cutting the breast and legs in half.

Stuffing a whole chicken before roasting adds extra flavour and plumps up the chicken. Do not stuff a chicken more than three hours before cooking. If using warm stuffing, the chicken must be cooked immediately. Stuffed, or stuffed and trussed chickens take a little longer to cook than unstuffed chickens. The juices from the cavity of the chicken will soak into the stuffing, so the stuffing must be cooked through to prevent contamination by harmful salmonella bacteria.

Some people prefer chicken without the skin. Removing the skin eliminates much of the fat from the chicken, as the fat lies in a layer just underneath the skin. Usually the skin is removed after cooking, but drumsticks can be skinned and then cooked. To remove the skin from drumsticks, use a small sharp knife. Begin by carefully loosening the skin from the flesh at the large joint end. Then pull the skin down and away from the flesh. Some chicken shops sell drumsticks with the skin already removed.

Boning a whole chicken is a technique used for special-occasion dishes. With bones removed, the chicken makes a meaty casing for a luxurious stuffing and the cooked chicken is much easier to carve or slice than a whole chicken. Chicken presented in this way is often served cold. Chicken wings can also be boned and stuffed, making an easy-to-eat appetizer or finger food for parties.

Marinating chicken gives it extra flavour and moisture. Marinades usually contain at least one acid ingredient, such as wine, vinegar, lemon juice or even yoghurt, to tenderize, plus other ingredients to flavour and colour the chicken.

Marinating is ideal to use in conjunction with quick-cooking methods. Frozen chicken should be thawed before marinating.

Place chicken in a shallow ceramic or glass (not metal) container and pour in the marinade. Stir well to make sure the chicken is well coated. Cover with plastic wrap and refrigerate, usually for at least two to eight hours, or preferably overnight, turning occasionally. When ready to cook, drain the chicken and reserve the liquid (or discard, according to the particular recipe) for basting during cooking and for making a sauce to serve with the finished dish, but make sure any sauces are boiled before serving. Honey or sugar should be used sparingly in marinades as sweet mixtures easily scorch during cooking.

COOKING TECHNIQUES

Chicken must always be eaten thoroughly cooked. To test when roasted, grilled or barbecued poultry is cooked, insert a skewer into the thickest part of the chicken (the thigh). If the juices run clear, the chicken is done. You can also test for doneness by twisting or jiggling the leg. If it moves easily in its socket, the chicken is cooked. Pan-fried, grilled or barbecued chicken is cooked when the meat is tender enough to fall easily off the bone when tested with a fork.

There are two basic cooking methods—using either dry heat or moist heat.

Dry heat cookery comprises oven roasting, barbecuing and grilling, stir-frying and pan-frying. With these methods, timing is important to prevent a tough, dry result.

Moist heat cookery comprises braising, casseroling, pot roasting, poaching and steaming. The less tender cuts are ideal here because they benefit from the long, slow cooking process.

Roasting: Stuff and truss the whole chicken if desired. Preheat the oven and have the chicken ready at room temperature. Use a shallow ovenproof pan that fits the chicken comfortably without squeezing it in. Place the chicken on a wire rack in the pan and pour a little wine or water into the bottom of the pan, if liked, to prevent the chicken from drying out. Brush the chicken all over with melted butter or oil. If the breast is browning too quickly, cover loosely with a piece of aluminium foil. Cook the chicken according to the directions in the recipe, basting occasionally. After roasting, let the chicken rest for 10 minutes, covered loosely with foil, before carving.

Barbecuing: Allow plenty of time for the barbecue to heat up, so that the chicken is cooking over glowing coals rather than flames. The cooking time of the chicken will depend on the thickness of the chicken pieces. There is a tendency for the outside of the chicken to cook too quickly when it is barbecued. If this happens, move the pieces further away from the heat, or brown the outside and then continue cooking, wrapped in aluminium foil. Serve the chicken wrapped in foil to preserve the juices. Brush any unwrapped pieces of chicken with marinade, butter or oil occasionally during cooking to prevent them from drying out.

To roast, place the chicken on a rack in a baking dish and brush with melted butter or oil.

If the juices run clear when a skewer is inserted into the thigh, the chicken is cooked.

To barbecue, place the chicken on a preheated grill and brush with marinade.

CRISPY CHICKEN WINGS

INGREDIENTS

12 chicken wings
3 tablespoons soy sauce
3 tablespoons hoisin sauce
125 g (1/2 cup) tomato sauce (ketchup)
2 tablespoons honey
1 tablespoon soft brown sugar
1 tablespoon cider vinegar
2 garlic cloves, crushed
1/4 teaspoon Chinese five-spice powder
2 teaspoons sesame oil

1 Tuck the chicken wing tips to the underside and place in a non-metallic bowl. Mix together all the remaining ingredients and pour over the wings, tossing to coat. Cover and leave in the fridge for at least 2 hours, turning occasionally. Drain, reserving the marinade.

2 Cook the wings on a hot, lightly oiled barbecue grill or flat plate for 5 minutes, or until cooked through, brushing with the reserved marinade several times.

SPICY BUFFALO WINGS WITH RANCH DRESSING

INGREDIENTS

12 large chicken wings
2 teaspoons garlic salt
2 teaspoons onion powder
oil, for deep-frying
125 ml (1/2 cup) tomato sauce (ketchup)
2 tablespoons Worcestershire sauce
50 g (1 3/4 oz) butter, melted
Tabasco sauce, to taste

Ranch dressing

1 small garlic clove, crushed
185 g (3/4 cup) mayonnaise
125 ml (1/2 cup) buttermilk
2 tablespoons finely chopped flat-leaf (Italian) parsley
1 tablespoon finely chopped chives
1 1/2 teaspoons lemon juice
1 1/2 teaspoons Dijon mustard
1 teaspoon onion powder

1 Pat the wings dry with paper towels, remove and discard the tip of each wing, then cut them in half at the joint. Combine the garlic salt, onion powder and 2 teaspoons of ground black pepper, and rub the spice mixture into each chicken piece.

2 Deep-fry the chicken in batches for 2–3 minutes without letting it brown, then remove from the oil and drain on crumpled paper towels. When the chicken has cooled a little, put it in a non-metallic bowl with the combined tomato sauce, Worcestershire sauce, butter and Tabasco, and toss so that all of the pieces are well coated in the marinade. Cover and refrigerate for at least 2 hours, or overnight.

3 To make the ranch dressing, mash the garlic and 1/4 teaspoon salt to a paste then add the mayonnaise, buttermilk, parsley, chives, lemon juice, mustard and onion powder, and whisk it all together. Season well, cover and chill for at least 1 hour before serving.

4 Preheat a barbecue to medium direct heat. Cook the chicken for 6–8 minutes on each side, or until it is caramelised and sticky, turning and basting with the marinade as it cooks. Serve hot with the ranch dressing.

CHICKEN BURGER WITH TARRAGON MAYONNAISE

INGREDIENTS

1 kg (2 lb) chicken mince
1 small onion, finely chopped
2 teaspoons lemon rind
2 tablespoons sour cream
1 cup (80 g/2¾ oz) fresh breadcrumbs
6 onion bread rolls

Tarragon mayonnaise
1 egg yolk
1 tablespoon tarragon vinegar
½ teaspoon French mustard
1 cup (250 ml/8 fl oz) olive oil

1. Place the chicken mince in a mixing bowl. Add the onion, lemon rind, sour cream and breadcrumbs. Using your hands, mix until thoroughly combined. Divide the mixture into 6 equal portions and shape into 1.5 cm (⅝ inch) thick patties.

2. Preheat a barbecue grill or flatplate to high. Lightly oil the hot barbecue. Cook the patties for 7 minutes each side, turning once. Serve the patties on the onion rolls with the mayonnaise.

3. To make the tarragon mayonnaise, place the egg yolk, half the vinegar and the mustard in a small mixing bowl. Whisk together for 1 minute, or until light and creamy. Add the oil about 1 teaspoon at a time, whisking constantly until the mixture thickens. Increase the flow of oil to a thin stream and continue whisking until all of the oil has been incorporated. Stir in the remaining vinegar and season with salt and white pepper.

HINT The mayonnaise can also be made in a food processor. Add the oil in a thin stream, with the motor constantly running, until the mixture thickens and turns creamy.

CHICKEN AND LIME HUMMUS TORTILLAS

INGREDIENTS

500 g (1 lb) chicken mince
1 red onion, finely chopped
3 cloves garlic, crushed
2 tablespoons chopped fresh mint
2 tablespoons chopped fresh parsley
2 tablespoons lime juice
2 eggs, lightly beaten
2 cups (160 g/5½ oz) fresh white breadcrumbs
2 teaspoons chicken stock powder
1½ cups (150 g/5 oz) dry breadcrumbs
oil, for shallow-frying
4 large flour tortillas
lettuce leaves, to serve
1 large ripe avocado, sliced

Lime Hummus
300 g (10 oz) can chickpeas, drained
2–3 tablespoons tahini
2 teaspoons sesame oil
2 cloves garlic, crushed
¼ cup (60 ml/2 fl oz) lime juice
1 tablespoon finely chopped fresh mint
½ teaspoon sweet paprika

1 In a bowl, combine the chicken mince, onion, garlic, herbs, lime juice, 1 egg fresh breadcrumbs, stock powder and some freshly ground black pepper. Use your hands to mix thoroughly. Shape 2 tablespoons of the mixture at a time into round patties dip in the remaining beaten egg, then toss in the dry breadcrumbs, pressing them on firmly.

2 Arrange the patties on a tray, cover and refrigerate for 30 minutes. Just before serving, shallow-fry the patties in moderately hot oil for 2–3 minutes each side, or until golden and cooked through. Drain on paper towels.

3 To make the lime hummus, process the chickpeas, tahini, sesame oil, garlic, lime juice and a little salt and pepper in a food processor until the mixture is a smooth, thick paste. Stir in the mint and paprika.

4 To serve, toast the tortillas under a grill or place in a dry pan until heated and lightly browned on both sides. Arrange lettuce leaves, a few chicken patties and some sliced avocado on each and top with the lime hummus.

CHICKEN BURGER WITH TANGY GARLIC MAYONNAISE

INGREDIENTS

4 chicken breast fillets
½ cup (125 ml/4 fl oz) lime juice
1 tablespoon sweet chilli sauce
4 slices bacon
4 hamburger buns, halved
4 lettuce leaves
1 large tomato, sliced

Garlic Mayonnaise
2 egg yolks
2 cloves garlic, crushed
1 tablespoon Dijon mustard
1 tablespoon lemon juice
½ cup (125 ml/4 fl oz) olive oil

1 Place the chicken in a shallow non-metal dish. Prick the chicken breasts with a skewer several times. Combine the lime juice and sweet chilli sauce in a jug. Pour the mixture over the chicken, cover and refrigerate for several hours or overnight.

2 To make the mayonnaise, place the egg yolks, garlic, mustard and lemon juice in a food processor bowl or blender and process until smooth. With the motor running, add the oil in a thin, steady stream. Process until the mixture reaches a thick consistency. Refrigerate, covered, until required.

3 Preheat a barbecue grill or flatplate to high. Remove and discard the rind from the bacon, and cut the bacon in half crossways. Lightly grease the hot barbecue. Cook the chicken and bacon for 5 minutes, or until crisp. Cook the chicken for a further 5–10 minutes, or until well browned and cooked through, turning once.

4 Toast the hamburger buns until lightly browned. Arrange the lettuce, tomato, chicken and bacon on the bases. Top with the garlic mayonnaise and finish with the remaining bun top.

CHICKEN FAJITAS

INGREDIENTS

4 chicken breast fillets
2 tablespoons olive oil
$^1/_4$ cup (60 ml) lime juice
2 cloves garlic, crushed
1 teaspoon ground cumin
$^1/_4$ cup (15 g/$^1/_2$ oz) chopped fresh coriander (cilantro) leaves
8 flour tortillas
1 tablespoon olive oil, extra
2 onions, sliced
2 green capsicums (peppers), cut into thin strips
1 cup (125 g/4 oz) grated Cheddar
1 large avocado, sliced
1 cup (250 g/8 oz) bottled tomato salsa

1 Trim the chicken of fat and sinew, and cut into thin strips. Place the chicken strips in a shallow non-metal dish. Combine the oil, lime juice, garlic, cumin and coriander in a jug, and mix well. Pour the mixture over the chicken. Refrigerate, covered, for several hours or overnight.

2 Preheat a barbecue grill or flatplate to high. Wrap the tortillas in foil and place on a cool part of the barbecue grill to warm through for 10 minutes. Heat the oil on a flatplate. Cook the onion and capsicum for 5 minutes, or until soft. Move the vegetables to a cooler part of the plate to keep warm.

3 Place the chicken and marinade on the flatplate and cook for 5 minutes, or until just tender. Transfer the chicken, vegetables and wrapped tortillas to a serving platter. Make up individual fajitas by placing the chicken, onion and capsicum, cheese and avocado over the tortillas. Top with the salsa and roll up to enclose the filling.

CHICKEN QUESADILLAS

INGREDIENTS

4 large green chillies
3 large red chillies
2 chicken breast fillets
3 tablespoons wholegrain mustard
2 tablespoons honey
4 flour tortillas
3 cups (375 g/12 oz) grated Cheddar
6 spring onions, thinly sliced
1–2 small red chillies, thinly sliced, optional
1–2 small green chillies, thinly sliced, optional
oilve oil, for cooking

Green Chilli Salsa

3 long green chillies, thinly sliced
2 tomatoes, peeled, seeded and chopped
1 onion, finely chopped
$^{1}/_{3}$ cup (10 g/$^{1}/_{4}$ oz) fresh coriander (cilantro) leaves, finely chopped
2 tablespoons lime juice

1 Place the large chillies under a hot grill and cook for 5–8 minutes, turning frequently, until the skins are blackened. Place the chillies in a plastic bag and leave to cool. Remove the skin and seeds. Cut them in half, then slice the flesh into thin strips.

2 Place the chicken in a deep dish. Combine the mustard and honey and add to the chicken. Turn the meat until well coated with the mixture. Cover and refrigerate for 1 hour. Place the chicken under a hot grill and cook for 4 minutes each side, or until tender. Cool slightly, then cut into thin strips.

3 To make the green chilli salsa, combine the chilli, tomato, onion, coriander and lime juice in a bowl and mix well, adding a little more lime juice if desired. Set aside to allow the flavours to combine.

4 Lightly grease a large, heavy-based frying pan and warm over medium heat. Place one tortilla in the pan and sprinkle with a quarter of the Cheddar and half the spring onion, roasted chilli, chicken and sliced chilli. Sprinkle with a third of the remaining Cheddar and top with another tortilla.

5 Brush lightly with a little oil, invert the quesadilla onto a plate and then slide back into the pan so that the top becomes the bottom. Cook for a few minutes longer, just until the Cheddar has melted and the underside looks golden and crisp. Slide the quesadilla onto a plate and keep warm.

6 Repeat with another tortilla, adding half of the remaining Cheddar, the remaining spring onion, roasted chilli, chicken and sliced chilli. Top with the rest of the Cheddar and the last tortilla. Serve immediately, cut into halves or triangles, with the salsa.

LOW-FAT CHICKEN PIES

INGREDIENTS

300 g (10 oz) chicken breast fillets
1 bay leaf
2 cups (500 ml/16 fl oz) chicken stock
2 large potatoes, chopped
250 g (8 oz) orange sweet potato (kumera), chopped
2 celery sticks, chopped
2 carrots, chopped
1 onion, chopped
1 parsnip, chopped
1 clove garlic, crushed
1 tablespoon cornflour (corn starch)
1 cup (250 ml/8 fl oz) skim milk
1 cup (155 g/5 oz) frozen peas, thawed
1 tablespoon chopped fresh chives
1 tablespoon chopped fresh parsley
$1\frac{1}{2}$ cups (185 g/6 oz) self-raising flour
20 g ($\frac{3}{4}$ oz) butter
$\frac{1}{3}$ cup (80 ml/$2\frac{3}{4}$ fl oz) milk
1 egg, lightly beaten

1 Combine the chicken, bay leaf and stock in a large, deep non-stick frying pan and simmer over low heat for about 10 minutes, until the chicken is cooked through. Remove the chicken, set aside and, when cool, cut into small pieces. Add the chopped potato, orange sweet potato, celery and carrot to the pan and simmer, covered, for about 10 minutes, or until just tender. Remove the vegetables from the pan with a slotted spoon.

2 Add the onion, parsnip and garlic to the pan and simmer, uncovered, for about 10 minutes, or until very soft. Discard the bay leaf. Purée the mixture in a food processor until smooth.

3 Stir the cornflour into 2 tablespoons of the skim milk until it forms a smooth paste. Stir the cornflour mixture into the puréed mixture with the remaining milk and then return to the pan. Stir over low heat until the mixture boils and thickens. Preheat the oven to moderately hot 200°C (400°F/Gas 6).

4 Combine the puréed mixture with the remaining vegetables, chicken, chives and parsley. Season with salt and pepper. Spoon into four $1\frac{3}{4}$ cup (440 ml/14 fl oz) ovenproof dishes.

5 To make the pastry, sift the flour into a large bowl, rub in the butter with your fingertips, then make a well in the centre. Combine the milk with $\frac{1}{3}$ cup (80 ml/$2\frac{3}{4}$ fl oz) water and add enough to the dry ingredients to make a soft dough. Turn out onto a lightly floured surface and knead until just smooth. Cut the dough into four portions and roll each out so that it is 1 cm ($\frac{1}{2}$ inch) larger than the top of the dish. Brush the edge of the dough with some of the egg and fit it over the top of each dish, pressing the edge firmly to seal.

6 Brush the pastry lightly with beaten egg and sprinkle with the sesame seeds. Bake for about 30 minutes, or until the tops are golden and the filling is heated through.

POTTED CHICKEN

INGREDIENTS

6 chicken thighs, skin removed
1 onion, sliced
1 carrot, sliced
6 peppercorns
1 bay leaf
pinch ground mace
pinch cayenne pepper
1/4 teaspoon freshly grated nutmeg
200 g (6 1/2 oz) unsalted butter, softened

1. Place the chicken thighs, onion, carrot, peppercorns and bay leaf in a pan and add 2 cups (500 ml/16 fl oz) of water. Bring to the boil, skimming off any foam. Reduce the heat, then cover and simmer for 30 minutes, or until the chicken is tender and cooked through.

2. Remove the chicken, then rapidly boil the remaining liquid until it has reduced to about 1/4 cup (60 ml/2 fl oz). Strain through a fine sieve and allow to cool.

3. Remove the chicken flesh from the bones. Place the flesh in a food processor with the liquid and process until smooth. Add the mace, cayenne pepper, nutmeg and 150 g (5 oz) of the butter. Season to taste with salt and pepper and process until combined and smooth.

4. Put the chicken mixture in a 3-cup (750 ml/24 fl oz) ceramic dish. Melt the remaining butter in a small pan and pour the yellow clarified butter onto the surface of the chicken, leaving the white milk solids in the pan. Refrigerate for 2 hours, or until the butter sets.

YAKITORI

INGREDIENTS

1 kg (2 lb) chicken thigh fillets
1/2 cup (125 ml/4 fl oz) sake
3/4 cup (185 ml/6 fl oz) shoyu (Japanese soy sauce)
1/2 cup (125 ml/4 fl oz) mirin
2 tablespoons sugar
10 spring onions, diagonally cut into 2 cm (3/4 inch) pieces

1 Soak 25 wooden skewers in water for about 20 minutes to prevent burning. Drain and set aside.

2 Cut the chicken thigh fillets into bite-sized pieces. Combine the sake, shoyu, mirin and sugar in a small pan. Bring the mixture to the boil and then set aside.

3 Thread the chicken pieces onto the wooden skewers alternately with the spring onion pieces. Place the chicken skewers on a foil-lined tray and cook them under a preheated grill, turning and brushing frequently with the sauce, for 7–8 minutes, or until the chicken is cooked through. Serve immediately, garnished with a few spring onion pieces.

CHICKEN AND TZATZIKI WRAP

INGREDIENTS

$^1/_2$ telegraph cucumber, seeded and grated
100 g (3$^1/_2$ oz) low-fat natural yoghurt
$^1/_4$ teaspoon lemon juice
1 tablespoon chopped mint
4 skinless chicken thigh fillets
pinch paprika
4 sheets lavash or other flat bread (see note)
4 large butter lettuce leaves

1 Sprinkle the grated cucumber with $^1/_2$ teaspoon salt. Leave the cucumber for 10 minutes, then drain and mix with the yoghurt, lemon juice and mint. Season.

2 Flatten the chicken thigh fillets with a meat mallet or rolling pin, season and sprinkle with the paprika. Grill the fillets for 5–7 minutes on each side, or until cooked through.

3 Lay out the lavash breads and place a large butter lettuce leaf on each. Spread each with one quarter of the tzatziki, then top with a sliced chicken fillet. Roll up, folding one end closed. Wrap in baking paper to serve.

NOTE If you can't find lavash, use any thin, flat bread that will roll up easily.

CHICKEN TIKKA

INGREDIENTS

1/4 onion, chopped
2 cloves garlic, crushed
1 tablespoon grated fresh ginger
2 tablespoons lemon juice
1 teaspoon grated lemon rind
3 teaspoons ground coriander (cilantro)
3 teaspoons ground cumin
3 teaspoons garam masala
1/3 cup (90 g/3 oz) plain yoghurt
1 teaspoon salt
750 g (1 1/2 lb) chicken thigh fillets, cut into cubes

1 Soak 10 wooden skewers in water for 30 minutes to prevent burning.

2 In a food processor, finely chop the onion, garlic, ginger, lemon juice, rind, coriander, cumin and garam masala. Stir in the yoghurt and salt.

3 Thread 4–5 chicken cubes onto each skewer and place in a arge shallow dish. Coat the skewers with the spice mixture. Marinate for several hours or overnight, covered, in the refrigerator.

4 Cook the skewers in batches on a barbecue or chargrill pan, or under a hot grill, for 3–4 minutes on each side, or until golden brown and cooked.

CHICKEN AND CORN PIES

INGREDIENTS

1 tablespoon olive oil
650 g (1 lb 7 oz) chicken thigh fillets, trimmed and cut into 1 cm (1/2 inch) pieces
1 tablespoon grated ginger
400 g (14 oz) oyster mushrooms, halved
3 corn cobs, kernels removed
125 ml (1/2 cup) chicken stock
2 tablespoons kecap manis
2 tablespoons cornflour (cornstarch)
30 g (1 cup) coriander (cilantro) leaves, chopped
6 sheets ready-rolled shortcrust pastry
milk, to glaze

1 Grease six metal pie tins measuring 9.5 cm (3 3/4 inches) on the base and 3 cm (1 1/4 inches) deep. Heat the oil in a large frying pan over high heat and add the chicken. Cook for 5 minutes, or until golden. Add the ginger, mushrooms and corn and cook for 5–6 minutes, or until the chicken is just cooked through. Add the stock and kecap manis.

2 Mix the cornflour with 2 tablespoons water in a small bowl or jug, then stir into the pan. Boil for 2 minutes before adding the coriander. Transfer to a bowl, cool a little, then refrigerate for 2 hours, or until cold.

3 Preheat the oven to moderate 180°C (350°F/Gas 4). Using a saucer to guide you, cut a 15 cm (6 inch) round from each sheet of shortcrust pastry and line the six pie tins. Fill the shells with the cooled filling, then cut out another six rounds large enough to make the lids. Top the pies with the lids, cut away any extra pastry and seal the edges with a fork. Decorate the pies with shapes cut from pastry scraps. Prick a few holes in the top of each pie, brush with a little milk and bake for 35 minutes, or until golden.

SATAY CHICKEN

INGREDIENTS

500 g (1 lb 2 oz) chicken thigh fillets, cut into 1 cm ($\frac{1}{2}$ inch) wide strips
1 garlic clove, crushed
2 teaspoons finely grated fresh ginger
3 teaspoons fish sauce

Satay sauce

2 teaspoons peanut oil
4 red Asian shallots, finely chopped
4 garlic cloves, crushed
2 teaspoons finely chopped fresh ginger
2 small red chillies, seeded and finely chopped
125 g ($\frac{1}{2}$ cup) crunchy peanut butter
185 ml ($\frac{3}{4}$ cup) coconut milk
2 teaspoons soy sauce
2 teaspoons grated palm sugar or soft brown sugar
$1\frac{1}{2}$ tablespoons fish sauce
1 fresh kaffir (makrut) lime leaf
$1\frac{1}{2}$ tablespoons lime juice

1 Put the chicken, garlic, ginger and fish sauce in a bowl and turn the chicken so that it is well coated. Cover the bowl and leave it in the refrigerator for 1 hour. Soak 12 wooden skewers in cold water for 1 hour.

2 To make the satay sauce, heat the oil in a saucepan over medium heat, then add the shallot, garlic, ginger and chilli. Stir the mixture constantly with a wooden spoon for 5 minutes, or until the shallots are golden. Reduce the heat to low, add the remaining sauce ingredients and simmer for 10 minutes, or until the sauce has thickened. Remove the lime leaf and keep the sauce warm while you cook the chicken.

3 Preheat the barbecue chargrill plate to medium–high direct heat. Thread two or three chicken strips onto each skewer, without crowding them, and grill the chicken for 10 minutes, or until it is cooked through, turning after 5 minutes. Serve the skewers with the satay sauce.

CHICKEN CURRY LAKSA

INGREDIENTS

Laksa paste

1 large onion, roughly chopped
5 cm (2 inch) piece fresh ginger, chopped
8 cm (3 inch) piece galangal, peeled and chopped
1 stem lemon grass, white part only, roughly chopped
2 cloves garlic
1 fresh red chilli, seeded and diced
2 teaspoons vegetable oil
2 tablespoons mild curry paste

500 g (1 lb) chicken breast fillets, cut into cubes
2 cups (500 ml/16 fl oz) chicken stock
60 g (2 oz) rice vermicelli
50 g (1$^3/_4$oz) dried egg noodles
400 ml (13 fl oz) light coconut milk
10 snow peas (mangetout), halved
3 spring onions, finely chopped
1 cup (90 g/3 oz) bean sprouts
$^1/_2$ cup (15 g/$^1/_2$ oz) fresh coriander (cilantro) leaves

1 To make the laksa paste, process the onion, ginger, galangal, lemon grass, garlic and chilli in a food processor until finely chopped. Add the oil and process until the mixture has a paste-like consistency. Spoon into a large wok, add the curry paste and stir over low heat for 1–2 minutes, until aromatic. Take care not to burn.

2 Increase the heat to medium, add the chicken and stir for 2 minutes, or until the chicken is well coated. Stir in the chicken stock and mix well. Bring slowly to the boil, then simmer for 10 minutes, or until the chicken is cooked through.

3 Meanwhile, cut the vermicelli into shorter lengths. Cook the vermicelli and egg noodles separately in large pans of boiling water for 5 minutes each. Drain and rinse in cold water.

4 Just prior to serving, add the light coconut milk and snow peas to the chicken and heat through. To serve, divide the vermicelli and noodles among four warmed serving bowls. Pour the hot laksa over the top and garnish with the spring onion, bean sprouts and coriander leaves.

HINT If you prefer a more fiery laksa, use a medium or hot brand of curry paste or increase the amount of chillies in your laksa paste.

CHINESE MUSHROOM AND CHICKEN SOUP

INGREDIENTS

3 dried Chinese mushrooms
185 g (6 oz) thin dried egg noodles
1 tablespoon oil
4 spring onions, julienned
1 tablespoon soy sauce
2 tablespoons rice wine, mirin or sherry (see note)
1.25 litres (5 1/4 cups) chicken stock
1/2 small barbecued chicken, shredded
50 g (1 3/4 oz) sliced ham, cut into strips
1 cup (90 g/3 oz) bean sprouts
fresh coriander (cilantro) leaves, to serve
thinly sliced red chilli, to serve

1. Soak the mushrooms in boiling water for 10 minutes to soften them. Squeeze dry then remove the tough stem from the mushrooms and slice them thinly.

2. Cook the noodles in a large pan of boiling water for 3 minutes, or according to the manufacturer's directions. Drain and cut the noodles into shorter lengths with scissors.

3. Heat the oil in a large heavy-based pan. Add the mushrooms and spring onion. Cook for 1 minute, then add the soy sauce, rice wine and stock. Bring slowly to the boil and cook for 1 minute. Reduce the heat then add the noodles, shredded chicken, ham and bean sprouts. Heat through for 2 minutes without allowing the soup to boil.

4. Use tongs to divide the noodles among four bowls, ladle in the remaining mixture, and garnish with coriander leaves and sliced chilli.

NOTE Rice wine and mirin are available at Asian food stores.

NEW POTATO, CHICKEN AND SPINACH FRITTATA

INGREDIENTS

600 g (1 1/4 lb) new potatoes, unpeeled, sliced about 1 cm (1/2 inch) thick
oil, for brushing
1 small barbecued chicken
500 g (1 lb) English spinach leaves, stalks removed
125 g (4 oz) feta, crumbled
1/2 cup (50 g/1 3/4 oz) grated Parmesan
1/2 cup (15 g/1/2 oz) basil leaves, chopped
10 eggs, lightly beaten

1. Preheat the oven to moderate 180°C (350°F/Gas 4). Put the potato in a greased ovenproof dish, brush all over with a little oil and bake for 30 minutes, or until cooked. Turn the potato once or twice during baking. Allow to cool.

2. Remove the skin from the chicken, pull the flesh from the bones and roughly chop the meat.

3. Wash the spinach, put in a large pan, cover and steam for 2 minutes to wilt slightly. Drain and allow to cool, then squeeze out the moisture and chop. Mix with the chicken and stir in the feta, Parmesan and basil. Season with salt and pepper, to taste.

4. Brush a 24 cm (9 1/2 inch) diameter non-stick pan with a little oil. Arrange half the potato slices over the base and spread half the chicken mixture on top, then repeat the layers. Season the beaten egg and pour over the top. Arrange the tomato quarters on top. Cook over low heat for 25 minutes, or until the centre is almost cooked. Take care not to burn the base. Heat the frittata under a preheated grill for 7 minutes, or until set. Cut the frittata into wedges to serve.

SMOKED CHICKEN CAESAR SALAD

INGREDIENTS

Garlic Croutons

1 thin baguette
45 g (1½ oz) unsalted butter
½ cup (125 ml/4 fl oz) olive oil
4 cloves garlic, crushed

1 cos lettuce, tough outer leaves discarded
1 large smoked chicken (about 950 g/1 lb 14 oz)
1½ cups (150 g/5 oz) Parmesan shavings

Dressing

2 eggs
2 cloves garlic, crushed
2 tablespoons lemon juice
2 teaspoons Dijon mustard
45 g (1½ oz) can anchovy fillets, drained
1 cup (250 ml/8 fl oz) olive oil
¼ teaspoon salt
1 teaspoon freshly ground black pepper

1. To make the garlic croutons, slice the baguette diagonally into 1 cm (½ inch) thick slices. Melt the butter and olive oil in a large frying pan over moderate heat. Stir in the crushed garlic. Fry the bread slices, a few at a time, until golden. Remove from the pan and drain on paper towels.

2. Separate the lettuce leaves, wash and dry thoroughly. Tear the larger leaves into pieces and refrigerate until well chilled. Cut the chicken into bite-sized chunks. Refrigerate while preparing the dressing.

3. To make the dressing, blend or process the eggs, garlic, lemon juice, mustard and anchovies. With the motor running, gradually pour in the oil in a thin stream and process until thick. Season with the salt and pepper.

4. In a large bowl, combine the torn lettuce leaves, chicken, about half of the croutons and half the Parmesan. Add the dressing and toss well. Arrange 2–3 whole lettuce leaves in each individual serving bowl, spoon in the salad and sprinkle with the remaining croutons and Parmesan. Season liberally with freshly ground black pepper and serve immediately.

THAI-STYLE CHICKEN AND BABY CORN SOUP

INGREDIENTS

150 g (5 oz) whole baby corn (see note)
1 tablespoon oil
2 stems lemon grass, white part only, very thinly sliced
2 tablespoons finely grated fresh ginger
6 spring onions, chopped
1 red chilli, finely chopped
4 cups (1 litre/34 fl oz) chicken stock
1 1/2 cups (375 ml/12 fl oz) coconut milk
250 g (8 oz) chicken breast fillets, thinly sliced
130 g (4 1/2 oz) creamed corn
1 tablespoon soy sauce
2 tablespoons finely chopped fresh chives, to serve
1 red chilli, thinly sliced, to serve

1. Cut the baby corn in half or quarters lengthways, depending on their size.
2. Heat the oil in a pan over medium heat. Cook the lemon grass, ginger, spring onion and chilli for 1 minute, stirring. Add the stock and coconut milk and bring to the boil—do not cover or the coconut milk will curdle.
3. Stir in the corn, chicken and creamed corn and simmer for 8 minutes, or until the corn and chicken are just tender. Add the soy sauce, season well and serve garnished with the chives and chilli.

TOM KHA GAI

INGREDIENTS

5 cm (2 inch) piece galangal (see hint)
2 cups (500 ml/16 fl oz) coconut milk
1 cup (250 ml/8 fl oz) chicken stock
3 chicken breast fillets, cut into thin strips
1–2 teaspoons finely chopped red chilli
2 tablespoons fish sauce
1 teaspoon soft brown sugar
1/4 cup (7 g/1/4 oz) fresh coriander (cilantro) leaves
fresh coriander (cilantro) sprigs, to serve

1 Peel the galangal and cut it into thin slices. Combine the galangal, coconut milk and stock in a medium pan. Bring to the boil and simmer, uncovered, over low heat for 10 minutes, stirring occasionally.

2 Add the chicken strips and chilli to the pan and simmer for 8 minutes.

3 Stir in the fish sauce and brown sugar. Add the coriander leaves and serve immediately, garnis

HINT If fresh galangal is not available, you can use 5 large slices of dried galangal instead. Prepare by soaking the slices in a little boiling water for 10 minutes and then cutting them into shreds. Add the soaking liquid to the chicken stock to make 1 cup (250 ml/8 fl oz) and use it in the recipe.

CHILLI CHICKEN AND CASHEW SALAD

INGREDIENTS

3 tablespoons sweet chilli sauce
2 tablespoons lime juice
2 teaspoons fish sauce
2 tablespoons chopped coriander (cilantro)
1 garlic clove, crushed
1 small red chilli, finely chopped
11/2 teaspoons grated fresh ginger
2 tablespoons olive oil
600 g chicken breast fillets
100 g salad leaves
250 g cherry tomatoes, halved
100 g Lebanese cucumber, cut into bite-sized chunks
50 g snow pea (mangetout) sprouts, trimmed
80 g (1/2 cup) cashew nuts, roughly chopped

1 Combine the chilli sauce, lime juice, fish sauce, coriander, garlic, chilli, ginger and 1 tablespoon of the oil in a large bowl.

2 Heat the remaining oil in a frying or chargrill pan over medium heat until hot, and cook the chicken for 5–8 minutes on each side or until cooked through. While still hot, slice each breast widthways into 1 cm (1/2 inch) slices and toss in the bowl with the dressing. Leave to cool slightly.

3 Combine the salad leaves, cherry tomatoes, cucumber chunks and snow pea sprouts in a serving bowl. Add the chicken and all of the dressing, and toss gently until the leaves are lightly coated. Scatter with chopped cashews and serve.

INDIAN MARINATED CHICKEN SALAD

INGREDIENTS

60 ml (1/4 cup) lemon juice
1 1/2 teaspoons garam masala
1 teaspoon ground turmeric
1 tablespoon finely grated fresh ginger
2 garlic cloves, finely chopped
3 1/2 tablespoons vegetable oil
3 chicken breast fillets (650 g/23 oz)
1 onion, thinly sliced
2 zucchini (courgettes), thinly sliced on the diagonal
100 g watercress leaves
150 g freshly shelled peas
2 ripe tomatoes, finely chopped
30 g (1 cup) coriander (cilantro) leaves

Dressing
1 teaspoon cumin seeds
1/2 teaspoon coriander (cilantro) seeds
90 g (1/3 cup) natural yoghurt
2 tablespoons chopped mint
2 tablespoons lemon juice

1. Combine the lemon juice, garam masala, turmeric, ginger, garlic and 2 teaspoons oil in a large bowl. Add the chicken fillets and onion, toss to coat in the marinade, cover, and refrigerate for 1 hour.

2. Remove and discard the onion then heat 2 tablespoons of oil in a large, frying pan. Cook the chicken for about 4–5 minutes on each side or until it is cooked through. Remove the chicken from the pan and leave for 5 minutes. Cut each breast across the grain into 1 cm (1/2 inch) slices.

3. Heat the remaining oil in the pan and cook the zucchini for 2 minutes, or until lightly golden and tender. Toss with the watercress in a large bowl. Cook the peas in boiling water for 5 minutes, or until tender, then drain. Rinse under cold water to cool. Add to the salad with the tomato, chicken and coriander.

4. For the dressing, gently roast the cumin and coriander seeds in a dry frying pan for 1–2 minutes, or until fragrant. Remove, then pound the seeds to a powder. Mix with the yoghurt, mint and lemon juice, then gently fold through the salad.

SMOKED CHICKEN AND PASTA SALAD WITH MUSTARD DRESSING

INGREDIENTS

1 tablespoon balsamic vinegar
150 ml (5 fl oz) olive oil
1 tablespoon lemon juice
3 tablespoons wholegrain mustard
200 g (7 oz) bucatini
450 g (16 oz) good-quality smoked chicken breast
8 small radishes
2 small Fuji apples
4 spring onions, finely sliced
35 g (1 oz) rocket

1 Combine the vinegar, olive oil, lemon juice and mustard in a screw-top jar, and shake well to combine. Season to taste with salt and pepper. Bring a large saucepan of salted water to the boil, and cook the bucatini according to the packet instructions until al dente. Drain, rinse under cold water and drain again. Toss one-third of the dressing through the bucatini and set aside for 30 minutes.

2 Cut the chicken breast on the diagonal and place in a large bowl. Thinly slice the radishes and add to the chicken. Quarter, core and cube the apples without peeling them, and add to the chicken with the sliced spring onion and rocket. Pour in the remaining dressing and toss lightly.

3 Gently mix the pasta with the smoked chicken until well combined. Divide the salad among four serving dishes and serve with fresh, crusty bread. May be served cold as a light meal.

NOTE Bucatini is a thick, spaghetti-like pasta with a hollow centre. It has a chewy texture. Smoked chicken often has a dark skin. You may wish to remove this to improve the appearance of the salad.

WARM CHICKEN AND PASTA SALAD

INGREDIENTS

375 g (13 oz) penne
100 ml (3 1/3 fl oz) olive oil
4 long, thin eggplants (aubergines), thinly sliced on the diagonal
2 chicken breast fillets
2 teaspoons lemon juice
15 g (1/2 cup) chopped flat-leaf parsley
270 g (9 1/2 oz) chargrilled red capsicum (pepper), drained and sliced (see note)
155 g (5 1/2 oz) fresh asparagus spears, trimmed, blanched and cut into 5 cm (2 inch) lengths
85 g (3 oz) semi-dried (sun-blushed) tomatoes, finely sliced
grated Parmesan cheese (optional)

1. Cook the pasta in a large saucepan of boiling water until al dente. Drain, return to the pan and keep warm. Heat 2 tablespoons of the oil in a large frying pan over high heat and cook the eggplant for 4–5 minutes, or until golden and cooked through.

2. Heat a lightly oiled chargrill pan over high heat and cook the chicken for 5 minutes each side, or until browned and cooked through. Cut into thick slices. Combine the lemon juice, parsley and the remaining oil in a small jar and shake well. Return the pasta to the heat, toss through the dressing, chicken, eggplant, capsicum, asparagus and tomato until well mixed and warmed through. Season with black pepper. Serve warm with a scattering of grated Parmesan, if desired.

NOTE Jars of chargrilled capsicum can be bought at the supermarket; otherwise, visit your local deli.

WARM MINTED CHICKEN AND PASTA SALAD

INGREDIENTS

250 g (9 oz) cotelli pasta
125 ml (1/2 cup) olive oil
1 large red capsicum (pepper)
3 chicken breast fillets
6 spring onions, cut into 2 cm (1 inch) lengths
4 garlic cloves, thinly sliced
35 g (3/4 cup) chopped mint
80 ml (1/3 cup) cider vinegar
100 g baby English spinach leaves

1. Cook the pasta in a large saucepan of boiling water until al dente, drain, stir in 1 tablespoon of the oil and set aside. Meanwhile, cut the capsicum into quarters, removing the seeds and membrane. Place, skin-side-up, under a hot grill for 8–10 minutes, or until the skin blackens and blisters. Cool in a plastic bag, then peel away the skin. Cut into thin strips. Place the chicken between two sheets of plastic wrap and press with the palm of your hand until slightly flattened.

2. Heat 1 tablespoon of the oil in a large frying pan, add the chicken and cook over medium heat for 2–3 minutes each side, or until light brown and cooked through. Remove from the pan and cut into 5 mm slices.

3. Add another tablespoon of the oil to the pan and add the spring onion, sliced garlic and capsicum and cook, stirring, for 2–3 minutes, or until starting to soften. Add 25 g (1/2 cup) of the mint, the vinegar and the remaining oil and stir until warmed through. In a large bowl, combine the pasta, chicken, spinach, onion mixture and remaining mint and toss well, seasoning to taste. Serve warm.

THAI CHICKEN WITH GLASS NOODLES

INGREDIENTS

4 tablespoons coconut cream
1 tablespoon fish sauce
1 tablespoon palm sugar
2 chicken breasts, skinned and cut into shreds
120 g (4½ oz) glass noodles
2 stems lemon grass
4 kaffir lime (makrut) leaves
1 red onion, finely chopped
a large handful coriander (cilantro) leaves, chopped
a large handful mint, chopped
1–2 red chillies, sliced
3 green bird's eye chillies, finely sliced
2 tablespoons roasted peanuts, chopped
1–2 limes, cut in halves or quarters

1 Mix the coconut cream in a small saucepan or a wok with the fish sauce and palm sugar and bring to the boil, then add the chicken and simmer until the chicken is cooked through. This should only take a minute if you stir it a couple of times. Leave the chicken to cool in the sauce. Soak the noodles in boiling water for a minute or two — they should turn translucent and soft when they are ready. Drain them, then, using a pair of scissors, cut them into shorter lengths.

2 Peel the lemon grass until you reach the first purplish ring, then trim off the root. Make two or three cuts down through the bulb-like root, finely slice across it until it starts to get harder, then throw the hard top piece away. Pull the stems out of the lime leaves by folding the leaves in half, with the shiny side inwards, and pulling down on the stalk. Roll up the leaves tightly, then slice them very finely across.

3 Put all the ingredients, except the lime, in a bowl with the noodles and chicken, with its sauce, and toss everything together. Now squeeze the lime pieces over the dish and toss again.

CHICKEN AND VEGETABLE SOUP

INGREDIENTS

1.5 kg (3 lb 5 oz) chicken
1 onion
2 large leeks, halved lengthwise and well washed
3 large celery stalks
5 black peppercorns
1 bay leaf
2 large carrots, peeled and diced
1 large swede (rutabaga), peeled and diced
2 large tomatoes, peeled, seeded and finely chopped
165 g (3/4 cup) barley
1 tablespoon tomato paste (purée)
2 tablespoons finely chopped flat-leaf (Italian) parsley

1. Put the chicken, onion, 1 leek, 1 celery stalk, halved, the peppercorns and bay leaf in a large saucepan and add enough water to cover. Bring to the boil, then reduce the heat and simmer for 1 1/2 hours, skimming any impurities that rise to the surface.

2. Strain the stock through a fine sieve and return to the cleaned saucepan. Discard the onion, leek, celery, peppercorns and bay leaf, and set the chicken aside. When it is cool enough to handle, discard the fat and bones, then shred the flesh, cover and chill.

3. Allow the stock to cool, then refrigerate overnight. Skim the fat from the surface, place the stock in a large saucepan and bring to the boil. Dice the remaining leek and celery and add to the soup with the carrot, swede, tomato, barley and tomato paste. Simmer for 45–50 minutes, or until the vegetables are cooked and the barley is tender. Stir in the parsley and shredded chicken. Simmer until warmed through and season.

CHICKEN AND PASTA SALAD

INGREDIENTS

250 g (8 oz) chicken breast fillet
$1\frac{1}{2}$ cups (375 ml/12 fl oz) chicken stock
350 g (11 oz) spiral pasta
150 g (5 oz) asparagus, cut into short lengths
150 g (5 oz) Gruyère cheese, grated
2 spring onions, finely sliced

Dressing

$\frac{1}{4}$ cup (60 ml/2 fl oz) olive oil
$\frac{1}{4}$ cup (60 ml/2 fl oz) lemon juice
$\frac{1}{2}$ teaspoon sugar

1 Put the chicken and stock in a frying pan. Bring to the boil, reduce the heat and poach gently, turning regularly, for 8 minutes, or until tender. Remove, cool and slice thinly.

2 Cook the pasta in a large pan of rapidly boiling salted water until al dente. Drain well and cool.

3 Cook the asparagus in boiling water for 2 minutes. Drain and place in a bowl of iced water. Drain again. Combine with the chicken, pasta and cheese in a large bowl.

4 To make the dressing, whisk the ingredients together. Season with salt and pepper. Add to the salad and toss well. Transfer to a serving bowl and scatter with the spring onions.

CHUNKY CHICKEN AND VEGETABLE SOUP

INGREDIENTS

1 tablespoon oil
1 carrot, sliced
1 leek, chopped
2 chicken thigh fillets, cut into bite-sized pieces
1/4 cup (35 g/1 oz) ditalini or other small pasta
1 litre (4 cups/34 fl oz) vegetable stock
2 ripe tomatoes, diced

1 Heat the oil in a saucepan and cook the carrot and leek over medium heat for 4 minutes, or until soft. Add the chicken and cook for a further 2 minutes, or until the chicken has changed colour.

2 Add the pasta and the vegetable stock, cover and bring to the boil. Reduce the heat and simmer for 10 minutes, or until the pasta is cooked. Add the tomato halfway through the cooking. Season to taste with salt and pepper. Serve with fresh crusty bread.

CREAMY CHICKEN AND PEPPERCORN PAPPADELLE

INGREDIENTS

2 chicken breast fillets
30 g (1 oz) butter
1 onion, halved and thinly sliced
2 tablespoons drained green peppercorns, slightly crushed
$^1/_2$ cup (125 ml/4 fl oz) white wine
300 ml (10 fl oz) cream
400 g (13 oz) pappardelle
90 g (3 oz) sour cream (optional)
2 tablespoons chopped fresh chives

1 Cut the chicken in half so that you have four flat fillets and season with salt and pepper. Melt the butter in a frying pan, add the chicken and cook for 3 minutes each side, or until lightly browned and cooked through. Remove from the pan, cut into slices and keep warm.

2 Add the onion and peppercorns to the same pan and cook over medium heat for 3 minutes, or until the onion has softened slightly. Add the wine and cook for 1 minute, or until reduced by half. Stir in the cream and cook for 4–5 minutes, or until thickened slightly, then season with salt and pepper.

3 Meanwhile, cook the pasta in a large pan of rapidly boiling salted water until al dente. Drain and return to the pan to keep warm. Mix together the pasta, chicken with any juices and cream sauce. Divide the pasta among serving bowls, top with a dollop of sour cream and sprinkle with chives.

SPICY CHICKEN BROTH WITH CORIANDER AND PASTA

INGREDIENTS

350 g (11 oz) chicken thighs or wings, skin removed
2 carrots, finely chopped
2 celery stalks, finely chopped
2 small leeks, finely chopped
3 egg whites
6 cups chicken stock
Tabasco sauce

Coriander Pasta

$^{1}/_{2}$ cup (60 g/2 oz) plain flour
1 egg
$^{1}/_{2}$ teaspoon sesame oil
small bunch fresh coriander (cilantro) leaves

1 Put the chicken pieces, carrot, celery and leek in a large heavy-based pan. Push the chicken to one side and add the egg whites to the vegetables. Using a wire whisk, beat for a minute or so until frothy (take care not to use a pan that can be scratched by the whisk).

2 Warm the stock in another pan, then add gradually to the first pan, whisking continuously to froth the egg whites. Continue whisking while slowly bringing to the boil. Make a hole in the froth on top with a spoon and leave to simmer, uncovered, for 30 minutes without stirring.

3 Line a large strainer with a damp tea towel or double thickness of muslin and strain the broth into a clean bowl (discard the chicken and vegetables). Season with salt, pepper and Tabasco to taste. Set aside.

4 To make the coriander pasta, sift the flour into a bowl and make a well in the centre. Whisk the egg and oil together and pour into the well. Mix together to make a soft pasta dough and knead on a lightly floured surface for 2 minutes, until smooth.

5 Divide the pasta dough into four even portions. Roll one portion out very thinly and cover with a layer of evenly spaced coriander leaves. Roll out another portion of pasta and lay this on top of the leaves, then gently roll the layers together. Repeat with the remaining pasta and coriander.

6 Cut out squares of pasta around the leaves. The pasta may then be left to sit and dry out if it is not needed immediately. When you are ready to serve, heat the chicken broth gently in a saucepan. As the broth simmers, add the pasta and cook for 1 minute. Serve immediately.

HINT Beg, borrow or steal a pasta machine for making this fine, delicate pasta. A rolling pin will suffice if necessary but you will need to roll the pasta as thinly as possible.

NOTE The egg whites added to the vegetable and chicken stock pot make the broth very clear rather than leaving it with the normal cloudy appearance of chicken stock. This is called clarifying the stock. When you strain the broth through muslin or a tea towel, don't press the solids to extract the extra liquid or the broth will become cloudy. It is necessary to make a hole in the froth on top to prevent the stock boiling over.

CHIANG MAI NOODLES

INGREDIENTS

250 g (9 oz) fresh thin egg noodles
2 tablespoons oil
6 red Asian shallots, finely chopped
3 cloves garlic, crushed
1–2 small fresh red chillies, seeded and finely chopped
2–3 tablespoons red curry paste
375 g (13 oz) chicken breast fillet, cut into thin strips
2 tablespoons fish sauce
1 tablespoon grated palm sugar
750 ml (3 cups/25 fl oz) coconut milk
1 tablespoon lime juice
250 ml (1 cup/8$^1/_2$ fl oz) chicken stock
4 spring onions, sliced, to garnish
$^1/_3$ cup (10 g/$^1/_3$ oz) fresh coriander (cilantro) leaves, to garnish
fried red Asian shallot flakes, to garnish
purchased fried noodles, to garnish
fresh red chilli, finely diced, to garnish

1. Cook the noodles in a saucepan of boiling water according to the packet instructions. Drain, cover and set aside.

2. Heat a large wok over high heat, add the oil and swirl to coat. Add the shallots, garlic and chilli, and stir-fry for 3 minutes. Stir in the curry paste and stir-fry for 2 minutes. Add the chicken and stir-fry for 3 minutes, or until it changes colour.

3. Stir in the fish sauce, palm sugar, coconut milk, lime juice and stock. Reduce the heat and simmer over low heat for 5 minutes—do not boil.

4. To serve, divide the noodles among four deep serving bowls and spoon in the chicken mixture. Garnish with the spring onion, coriander, shallot flakes, noodles and chilli.

CHICKEN WITH PONZU SAUCE AND SOMEN NOODLES

INGREDIENTS

Ponzu sauce
1 tablespoon lemon juice
1 tablespoon lime juice
1 tablespoon rice vinegar
1 tablespoon tamari
1½ tablespoons mirin
2½ tablespoons Japanese soy sauce
5 cm (2 inch) piece kombu (kelp), wiped with a damp cloth
1 tablespoon bonito flakes

900 g (2 lb) chicken thighs, trimmed and cut in half across the bone
10 cm (4 inch) piece kombu (kelp)
200 g (7 oz) dried somen noodles
250 g (9 oz) shiitake mushrooms (cut into smaller pieces if too large)
1 carrot, thinly sliced
300 g (10½ oz) baby English spinach leaves

1 To make the sauce, combine all the ingredients in a non-metallic bowl. Cover with plastic wrap and refrigerate overnight, then strain through a fine sieve.

2 Put the chicken and kombu in a pan with 875 ml (3½ cups) water. Bring to a simmer over medium heat and cook for 20 minutes, or until the chicken is cooked, skimming the scum off the surface. Remove the chicken and strain the broth. Transfer the broth and chicken pieces to a 2.5 litre (10 cup) flameproof casserole dish or Japanese nabe. Cover and continue to cook over low heat for 15 minutes.

3 Meanwhile, cook the noodles in a large saucepan of boiling water for 2 minutes, or until tender. Drain and rinse under cold running water.

4 Add the mushrooms and carrot to the chicken and cook for 5 minutes. Put the noodles on top of the chicken, then top with the spinach. Cover and cook for 2 minutes, or until the spinach has just wilted. Stir in 4–6 tablespoons of the ponzu sauce and serve.

NOTE Traditionally, this dish would be served in a ceramic nabe dish, for the guests to help themselves.

BACON-WRAPPED CHICKEN

INGREDIENTS

2 tablespoons olive oil
2 tablespoons lime juice
$^1/_4$ teaspoon ground coriander (cilantro)
6 chicken breast fillets
4 tablespoons fruit chutney
3 tablespoons chopped pecan nuts
6 rashers bacon

1 Mix together the olive oil, lime juice, coriander and salt and pepper. Using a sharp knife, cut a pocket in the thickest section of each fillet. Mix together the chutney and nuts. Spoon 1 tablespoon of the chutney mixture into each chicken breast pocket.

2 Turn the tapered ends of the fillets to the underside. Wrap a slice of bacon around each fillet to enclose the filling and secure with a toothpick.

3 Put the chicken parcels on a hot, lightly oiled barbecue grill or flat plate and cook for 5 minutes on each side, or until cooked through, turning once. Brush with the lime juice mixture several times during cooking and drizzle with any leftover lime juice mixture to serve.

NOTE This recipe also works well with prosciutto, which is an Italian equivalent of bacon.

FIVE SPICE ROAST CHICKEN

INGREDIENTS

1.8 kg (4 lb) chicken
1 tablespoon soy sauce
2 garlic cloves, crushed
1 teaspoon finely grated fresh ginger
1 tablespoon honey
1 tablespoon rice wine
1 teaspoon five-spice
1 tablespoon peanut oil

1 Wash the chicken and pat it thoroughly dry inside and out with paper towels. Whisk the soy sauce, garlic, ginger, honey, rice wine and five-spice together in a small bowl and brush it all over the chicken, ensuring every bit of skin is well coated. Put the chicken on a wire rack over a baking tray and refrigerate it, uncovered, for at least 8 hours or overnight.

2 Preheat a kettle or covered barbecue to medium indirect heat and put a drip tray under the rack. Brush the chicken liberally with the peanut oil and put it breast-side up in the middle of the barbecue over the drip tray. Cover the barbecue and roast the chicken for 1 hour 10 minutes, or until the juices run clear when you pierce it with a skewer between the thigh and body. Check the chicken every so often, and if it appears to be over-browning, cover it loosely with foil. Leave it to rest, covered, for 10 minutes before carving and serving. The flavours in this style of chicken go particularly well with steamed Asian greens and fried rice.

HONEY MUSTARD CHICKEN

INGREDIENTS

175 g (1/2 cup) honey
60 g (1/4 cup) Dijon mustard
2 tablespoons oil
2 tablespoons white wine vinegar
3 garlic cloves, crushed
2 tablespoons chopped parsley leaves
1.8 kg (4 lb) chicken, cut into 10 serving pieces

1 Put the honey, mustard, oil, white wine vinegar, garlic, parsley and 1/4 teaspoon freshly ground black pepper in a large, non-metallic bowl. Mix it all together well, and put aside 60 ml (1/4 cup) of the marinade to baste the chicken during cooking. Add the chicken pieces to the rest of the marinade and turn them so that they are thoroughly coated. Cover the bowl and refrigerate it for at least 4 hours, or overnight.

2 Preheat a covered or kettle barbecue to medium indirect heat and cook the chicken pieces for 20–30 minutes, or until they are cooked through. The breast pieces may take as little as 15 minutes, while dark meat will take longer. Baste the chicken with the reserved marinade during the last 5–8 minutes of cooking, but no earlier or it is likely to burn. The chicken is delicious served on a bed of spring onion mash.

LEBANESE CHICKEN

INGREDIENTS

250 g (1 cup) plain Greek-style yoghurt
2 teaspoons soft brown sugar
4 garlic cloves, crushed
3 teaspoons ground cumin
$1^{1}/_{2}$ teaspoons ground coriander (cilantro)
7 g ($^{1}/_{4}$ cup) chopped flat-leaf (Italian) parsley
60 ml ($^{1}/_{4}$ cup) lemon juice
1 x 1.8 kg (4 lb) chicken, cut into 10 serving pieces
cooking oil spray

1. Put the yoghurt, brown sugar, garlic, cumin, coriander, chopped parsley and lemon juice in a large non-metallic bowl and mix them together. Add the chicken pieces to the marinade and turn them so that they are completely coated, then cover and refrigerate for at least 2 hours, or overnight.

2. Lightly spray the barbecue plates with oil, then preheat the barbecue to medium direct heat. Take the chicken pieces out of the marinade and season them with salt and pepper. Cook the chicken pieces on the flat plate, turning them frequently, for 20–30 minutes, or until they are cooked through. If you have a barbecue with a lid, cover the barbecue while the chicken is cooking. This way, the breast pieces will take only 15 minutes to cook, while the pieces on the bone will take about 10 minutes longer.

MARGARITA CHICKEN

INGREDIENTS

4 chicken breasts, skin on, tenderloin and any excess fat removed
60 ml (1/4 cup) tequila
60 ml (1/4 cup) lime juice
2 small chillies, finely chopped
3 garlic cloves, crushed
15 g (1/4 cup) finely chopped coriander (cilantro) leaves
1 tablespoon olive oil
lime wedges

1 Put the chicken, tequila, lime juice, chilli, garlic, coriander and olive oil in a non-metallic bowl and mix it all together so that the chicken is coated in the marinade. Cover the bowl and refrigerate for at least 2 hours, or preferably overnight.

2 Preheat a barbecue chargrill to medium–high direct heat. Remove the chicken breasts from the marinade, season them with salt and pepper, and grill for 7–8 minutes on each side or until they are cooked through.

3 Slice the chicken breasts on the diagonal and serve with lime wedges.

MIRIN AND SAKE CHICKEN

INGREDIENTS

4 large chicken breast fillets
2 tablespoons mirin
2 tablespoons sake
1 tablespoon oil
5 cm (2 inch) piece of fresh ginger, very finely sliced
3 teaspoons soy sauce
salad leaves, to serve

1. Put the chicken in a non-metallic dish. Combine the mirin, sake and oil and pour over the chicken. Marinate for 15 minutes, then drain the chicken, reserving the marinade.

2. Cook the chicken on a hot, lightly oiled barbecue grill or flat plate for 4 minutes each side, or until tender.

3. Put the ginger in a pan and add the reserved marinade. Boil for about 7 minutes, or until thickened. Drizzle the soy sauce over the chicken and top with the ginger. Serve immediately on a bed of salad leaves.

SAGE AND RICOTTA STUFFED CHICKEN

INGREDIENTS

250 g (1 cup) fresh ricotta cheese, well drained
1 tablespoon shredded sage leaves
2 garlic cloves, crushed
1 1/2 teaspoons grated lemon zest
40 g (1 1/2 oz) finely grated Parmesan cheese
4 chicken breast fillets, tenderloin removed
8 thin slices prosciutto
olive oil, for brushing

1 Mix together the ricotta, sage, garlic, zest and Parmesan until they are well combined. Use a sharp knife to cut a large pocket into the side of each chicken breast and fill each pocket with a quarter of the ricotta mixture. Pin the pockets closed with toothpicks and wrap each breast in two slices of prosciutto, securing it with a toothpick.

2 Heat a barbecue flat plate to medium direct heat, brush the chicken parcels with olive oil and season them with freshly ground black pepper. Cook them for 8 minutes on each side, or until they are cooked through. This is delicious served with baby spinach salad.

BRAISED CHICKEN AND LEEK IN WINE

INGREDIENTS

2 tablespoons oil
1.2 kg (2 lb 6½ oz) chicken pieces
1 leek, thinly sliced
5 spring onions, thinly sliced on the diagonal
2 tablespoons fresh marjoram
150 ml (5 fl oz) white wine
400 ml (13 fl oz) chicken stock
100 ml (3½ fl oz) cream

1. Score the chicken drumsticks. Heat 1 tablespoon of the oil in a frying pan and cook the chicken pieces in batches for 3–4 minutes, or until browned.

2. Heat the remaining oil in a large casserole dish and cook the leek, spring onion and marjoram for 4 minutes, or until soft. Add the chicken and wine, and cook for 2 minutes. Add the stock, cover and bring to the boil. Reduce the heat and simmer for 30 minutes. Stir in the cream and simmer, uncovered, for 15 minutes, or until the chicken is tender. Season to taste. Serve with steamed rice.

THAI SPICED CHICKEN WITH POTATO ROSTI

INGREDIENTS

600 g (1 lb 5 oz) chicken breast fillet, cut into strips
1 tablespoon chopped lemon grass
2 tablespoons lime juice
1$^1/_2$ tablespoons oil
2 garlic cloves, crushed
1 tablespoon grated fresh ginger
2 teaspoons sweet chilli sauce
2 spring onions (scallions), chopped
1 lime, cut into 6 wedges

Potato rosti
600 g (1 lb 5 oz) potatoes
3 tablespoons plain (all-purpose) flour
1 egg, lightly beaten

1 Remove any excess fat or sinew from the chicken and put the chicken in a shallow, non-metallic dish. Mix together the lemon grass, lime juice, oil, garlic, ginger, sweet chilli sauce and spring onion. Pour over the chicken pieces, cover and refrigerate for at least 2 hours.

2 To make the potato rosti, peel and grate the potatoes. Squeeze the excess moisture from the potato with your hands until it feels quite dry. Mix the potato with the flour and egg and season well. Divide into six equal portions. Cook on a hot, lightly oiled barbecue flat plate for 10 minutes, or until golden brown on both sides, flattening them down with the back of a spatula during cooking.

3 Drain the chicken and reserve the marinade. Cook on a barbecue grill or flat plate for 3 minutes each side, or until tender and golden brown. Brush with the reserved marinade while cooking. Serve with the rosti and a squeeze of lime juice.

CHICKEN AND CIDER STEW WITH APPLE AND POTATO MASH

INGREDIENTS

1 kg (2 lb) chicken thigh fillets, cut into 2 cm (¾ inch) cubes
1½ tablespoons finely chopped fresh thyme
1 tablespoon oil
90 g (3 oz) butter
3 French shallots, thinly sliced
1½ cups (375 ml/12 fl oz) apple cider
1 kg (2 lb) potatoes, cubed
2 large green apples, peeled, cored and sliced into eighths
⅔ cup (170 ml/5½ fl oz) cream

1 Season the chicken thighs with 2 teaspoons of the thyme and salt and black pepper. Heat the oil and 20 g (¾ oz) of the butter in a large saucepan over medium–high heat. Cook the chicken in two batches for 2–3 minutes, or until evenly browned. Remove from the pan.

2 Add the French shallots and the remaining thyme to the pan, and sauté for 2 minutes. Pour in the cider, then bring to the boil, scraping off any sediment that has stuck to the bottom of the pan. Return the chicken to the pan and cover. Reduce the heat to medium–low and cook for 35–40 minutes, or until the chicken is tender and the sauce has reduced (check occasionally to see if any water needs to be added).

3 Meanwhile, cook the potato and apple in a saucepan of boiling water for 15–20 minutes, or until tender. Drain and return to the pan over low heat for 1 minute to allow any water to evaporate. Remove from the heat, and mash with a potato masher. Stir in 2 tablespoons of the cream and the remaining butter with a wooden spoon, then season with salt and pepper.

4 Gently stir the remaining cream into the chicken stew and cook for a further 2–4 minutes, or until the sauce has thickened. Serve at once with the potato and apple mash and a crisp green salad.

CHICKEN, ARTICHOKE AND BROAD BEAN STEW

INGREDIENTS

1 cup (155 g/5 oz) frozen broad beans
8 chicken thighs (skin removed, optional)
$^1/_2$ cup (60 g/2 oz) seasoned plain flour
2 tablespoons oil
1 large red onion, cut into small wedges
$^1/_2$ cup (125 ml/4 fl oz) dry white wine
1$^1/_4$ cups (310 ml/10 fl oz) chicken stock
2 teaspoons finely chopped fresh rosemary
340 g (11 oz) marinated artichokes, well drained and quartered
800 g (1 lb 10 oz) potatoes, cut into large cubes
60 g (2 oz) butter

1 Remove the skins from the broad beans. Coat the chicken in the flour, shaking off the excess. Heat the oil in a saucepan or flameproof casserole dish, then brown the chicken in two batches on all sides over medium heat. Remove and drain on paper towels.

2 Add the onion to the pan and cook for 3–4 minutes, or until soft but not brown. Increase the heat to high, pour in the wine and boil for 2 minutes, or until reduced to a syrup. Stir in 1 cup (250 ml/8 fl oz) of the stock and bring just to the boil, then return the chicken to the pan with the rosemary. Reduce the heat to low and simmer, covered, for 45 minutes.

3 Add the artichokes to the pan, increase the heat to high and return to the boil. Reduce to a simmer and cook, uncovered, for 10–15 minutes. Add the beans and cook for a further 5 minutes.

4 Meanwhile, cook the potato in a saucepan of boiling water for 15–20 minutes, or until tender. Drain, then return to the pan. Add the butter and the remaining stock, and mash with a potato masher. Serve on the side of the stew.

MOROCCAN CHICKEN

INGREDIENTS

8 large chicken drumsticks
3 cloves garlic, crushed
1 teaspoon grated fresh ginger
1 teaspoon ground turmeric
2 teaspoons ground cumin
1 teaspoon ground cardamom
1 teaspoon finely grated lemon rind
2 tablespoons oil
1 onion, sliced
2 cups (500 ml/16 fl oz) chicken stock
6 pitted dates, chopped
1/3 cup (20 g/3/4 oz) shredded coconut

1 Trim the chicken of excess fat and sinew. Place the chicken in a large bowl. Combine the garlic, ginger, turmeric, cumin, cardamom and rind in a small bowl. Add to the chicken and stir to completely coat. Cover and marinate for 2 hours.

2 Preheat the oven to moderate 180°C (350°F/Gas 4). Heat the oil in a large heavy-based frying pan. Cook the chicken quickly over medium heat until well browned. Drain on paper towels. Place the chicken in an ovenproof casserole dish.

3 Add the onion to the pan and cook, stirring, for 5 minutes, or until soft. Add the cooked onion, chicken stock, dates and shredded coconut to the casserole dish. Cover and bake for 1 hour 15 minutes, or until the chicken is tender, stirring occasionally

NOTE The chicken may be left to marinate overnight in the refrigerator.

GREEN CHICKEN CURRY

INGREDIENTS

2 cups (500 ml/16 fl oz) coconut cream (do not shake the can—see note)
4 tablespoons green curry paste
2 tablespoons grated palm sugar or soft brown sugar
2 tablespoons fish sauce
4 kaffir lime (makrut) leaves, finely shredded
1 kg (2 lb) chicken thigh or breast fillets, cut into thick strips
200 g ($6\frac{1}{2}$ oz) bamboo shoots, cut into thick strips
100 g ($3\frac{1}{2}$ oz) snake beans, cut into 5 cm (2 inch) lengths
$\frac{1}{2}$ cup (15 g/$\frac{1}{2}$ oz) fresh Thai basil leaves

1 Place $\frac{1}{2}$ cup (125 ml/4 fl oz) of the thick coconut cream from the top of the can in a wok, and bring to the boil. Add the curry paste, then reduce the heat and simmer for 15 minutes, or until fragrant and the oil starts to separate from the cream. Add the palm sugar, fish sauce and kaffir lime leaves to the pan.

2 Stir in the remaining coconut cream and the chicken, bamboo shoots and beans, and simmer for 15 minutes, or until the chicken is tender. Stir in the Thai basil and serve with rice.

NOTE Do not shake the can of coconut cream because good-quality coconut cream has a layer of very thick cream at the top that has separated from the rest of the cream. This has a higher fat content, which causes it to split or separate more readily than the rest of the coconut cream or milk.

PONZU CHICKEN AND NOODLE HOTPOT

INGREDIENTS

Ponzu Sauce

1 tablespoon lemon juice
1 tablespoon lime juice
1 tablespoon rice vinegar
1 tablespoon tamari
1½ tablespoons mirin
2½ tablespoons Japanese soy sauce
5 cm (2 inch) piece kombu (kelp), wiped with a damp cloth
1 tablespoon bonito flakes

900 g (1 lb 13 oz) chicken thigh cutlets, cut in half across the bone
10 cm (4 inch) piece kombu (kelp)
200 g (6½ oz) dried somen noodles
250 g (8 oz) fresh shiitake mushrooms, halved if large
1 carrot, thinly sliced
300 g (10 oz) baby English spinach leaves

1 To make the ponzu sauce, combine all the ingredients in a non-metallic bowl. Cover with plastic wrap and refrigerate overnight, then strain through a fine sieve.

2 Place the chicken and kombu in a large saucepan with 3½ cups (875 ml/28 fl oz) water. Bring to a simmer over medium heat. Cook for 20 minutes, or until the chicken is cooked, skimming the scum off the surface. Remove the chicken pieces and strain the broth. Transfer the broth and chicken pieces to a 2.5 litre (4½ pint) flameproof casserole dish or Japanese nabe. Cover and continue to cook over low heat for 15 minutes.

3 Meanwhile, cook the noodles in a large saucepan of boiling water for 2 minutes, or until tender. Drain and rinse under cold running water.

4 Add the mushrooms and carrot to the chicken, and cook for 5 minutes. Place the noodles on top of the chicken, then top with the spinach. Cook, covered, for 2 minutes, or until the spinach has just wilted. Stir in 4–6 tablespoons of the ponzu sauce, or to taste. Serve immediately.

NOTE Traditionally, this dish would be served in a ceramic nabe dish, for your guests to help themselves.

SICHUAN CHICKEN

INGREDIENTS

1/4 teaspoon five-spice powder
750 g (1 1/2 lb) chicken thigh fillets, halved
2 tablespoons peanut oil
1 tablespoon julienned fresh ginger
1 teaspoon Sichuan peppercorns, crushed
1 teaspoon chilli bean paste (toban djan)
2 tablespoons light soy sauce
1 tablespoon Chinese rice wine
1 1/4 cups (250 g/8 oz) jasmine rice
600 g (1 1/4 lb) baby bok choy, leaves separated

1. Sprinkle the five-spice powder over the halved chicken fillets. Heat a wok until very hot, add half the oil and swirl to coat the side. Add the chicken pieces and cook for 2 minutes each side, or until browned. Remove from the wok.

2. Reduce the heat to medium. Add the julienned ginger to the wok and cook for 30 seconds. Add the crushed Sichuan peppercorns and chilli bean paste. Return the chicken pieces to the wok, add the soy sauce, wine and 1/2 cup (125 ml/4 fl oz) water, then simmer for 15–20 minutes, or until the chicken is cooked through.

3. Meanwhile, bring a large saucepan of water to the boil. Add the rice and cook for 12 minutes, stirring occasionally. Drain well.

4. Heat the remaining peanut oil in a saucepan. Add the baby bok choy and toss gently for 1 minute, or until the leaves wilt and the stems are tender. Serve with the chicken and boiled rice.

SPICY APRICOT CHICKEN

INGREDIENTS

750 g (1 1/2 lb) chicken thigh fillets, cut into 5 cm (2 inch) pieces
plain flour, for coating
2 tablespoons oil
2 teaspoons red curry paste
2 spring onions, sliced
415 g (13 oz) can apricot halves in light syrup
1/2 cup (125 ml/4 fl oz) chicken stock
200 g (6 1/2 oz) plain yoghurt
2 tablespoons chopped fresh coriander (cilantro) leaves
fresh coriander (cilantro) leaves, extra, to garnish

1. Lightly coat the chicken in the flour. Heat the oil in a saucepan, add the curry paste and stir over low heat for 1 minute. Add the spring onion and chicken, and cook, stirring, over medium heat for 2–3 minutes, or until the chicken is golden.

2. Drain the apricots and reserve 1/2 cup (125 ml/4 fl oz) apricot juice. Add the reserved juice, apricots and stock to the pan. Bring to the boil, then reduce the heat and simmer for 10 minutes, or until the chicken is tender.

3. Mix together the yoghurt and coriander, and place a spoonful of the mixture over each serving of chicken. Garnish with extra coriander leaves. Serve with couscous or rice.

VIETNAMESE CHICKEN CURRY

INGREDIENTS

4 large chicken Marylands (leg and thigh pieces)
1 tablespoon good-quality Indian curry powder
1 teaspoon caster sugar
$^{1}/_{3}$ cup (80 ml/2$^{3}/_{4}$ fl oz) oil
500 g (1 lb) orange sweet potato, cut into 3 cm (1$^{1}/_{4}$ inch) cubes
1 large onion, cut into thin wedges
4 cloves garlic, chopped
1 stem lemon grass (white part only), finely chopped
2 bay leaves
1 large carrot, cut into 1 cm ($^{1}/_{2}$ inch) pieces on the diagonal
400 ml (13 fl oz) can coconut milk

1 Remove the skin and any excess fat from the chicken. Pat dry with paper towels and cut each piece into 3 even pieces, making 12 pieces. Place the curry powder, sugar, $^{1}/_{2}$ teaspoon black pepper and 2 teaspoons salt in a bowl, and mix well. Rub the curry mixture into the chicken pieces. Place the chicken on a plate, cover with plastic wrap and refrigerate overnight.

2 Heat the oil in a large saucepan. Add the sweet potato cubes and cook over medium heat for 3 minutes, or until lightly golden. Remove with a slotted spoon.

3 Remove all but 2 tablespoons of the oil from the pan. Add the onion and cook, stirring, for 5 minutes. Then add the garlic, lemon grass and bay leaves, and cook for 2 minutes.

4 Add the chicken and cook, stirring, over medium heat for 5 minutes, or until the chicken is well coated in the mixture and starting to change colour. Add 1 cup (250 ml/8 fl oz) water and simmer, covered, for 20 minutes. Stir once or twice.

5 Stir in the carrot, sweet potato and coconut milk, and simmer, uncovered, stirring occasionally, for 30 minutes, or until the chicken is cooked and tender. Be careful not to break up the sweet potato cubes. Serve with steamed rice or rice stick noodles.

CREAMY TOMATO AND CHICKEN STEW

INGREDIENTS

4 slices bacon
2 tablespoons oil
50 g (1³/₄ oz) butter
300 g (10 oz) small button mushrooms, halved
1.5 kg (3 lb) chicken pieces
2 onions, chopped
2 cloves garlic, crushed
400 g (13 oz) can tomatoes
1 cup (250 ml/8 fl oz) chicken stock
1 cup (250 ml/8 fl oz) cream
2 tablespoons chopped fresh parsley
2 tablespoons fresh lemon thyme leaves

1 Chop the bacon into large pieces. Place a large, heavy-based pan over medium heat. Brown the bacon, then remove and set aside on paper towels.

2 Heat half the oil and a third of the butter in the pan until foaming, then stir in the mushrooms and cook until softened and golden brown. Remove from the pan with a slotted spoon.

3 Add the remaining oil to the pan with a little more butter. When the oil is hot, brown the chicken pieces in batches over high heat until the skin is golden all over and a little crisp. Remove from the pan.

4 Heat the remaining butter in the pan. Add the onion and garlic and cook over medium-high heat for about 3 minutes, or until softened. Pour in the tomatoes, stock and cream. Return the bacon, mushrooms and chicken pieces to the pan and simmer over medium-low heat for 25 minutes. Stir in the herbs, season with salt and freshly ground pepper and simmer for another 5 minutes before serving.

MEXICAN CHICKEN CASSEROLE

INGREDIENTS

$^{3}/_{4}$ cup (165 g/5$^{1}/_{2}$ oz) short-grain rice
300 g (10 oz) can red kidney beans, drained and thoroughly rinsed
3 $^{1}/_{2}$ tablespoons chopped fresh coriander (cilantro) leaves
1 tablespoon oil
600 g (1$^{1}/_{4}$ lb) chicken thigh fillets
2 x 200 g (6 $^{1}/_{2}$ oz) jars spicy taco sauce
2 cups (250 g/8 oz) grated Cheddar
$^{1}/_{2}$ cup (125 g/4 oz) sour cream

1. Preheat the oven to moderate 180°C (350°F/Gas 4). Lightly grease a 7 cm (2$^{3}/_{4}$ inch) deep, 21 cm (8$^{1}/_{2}$ inch) round casserole dish. Bring a large saucepan of water to the boil, add the rice and cook for 10–12 minutes, stirring occasionally. Drain.

2. In the prepared dish, combine the beans and 1$^{1}/_{2}$ tablespoons of the coriander, then add the rice and toss together. Lightly press the mixture so the beans are mixed into the rice and the mixture is flat.

3. Heat the oil in a large frying pan over medium–high heat. Sauté the chicken thighs for 3 minutes, then turn over. Add the spicy taco sauce, and cook for a further 3 minutes.

4. To assemble, spread half the cheese over the rice. Arrange the chicken and sauce on top in a star shape, sprinkle with 1$^{1}/_{2}$ tablespoons of the coriander, then sprinkle with the remaining cheese. Cover with foil.

5. Bake for 35–40 minutes, or until the mixture is bubbling and the cheese is melted and slightly browned—remove the foil for the last 5 minutes of cooking. Cut into four servings with a knife and scoop out carefully, keeping the layers intact. Serve sprinkled with the remaining coriander and a dollop of sour cream.

CHICKEN KAPITAN

INGREDIENTS

1 teaspoon small dried shrimp
1/3 cup (80 ml/2 3/4 fl oz) oil
6–8 fresh red chillies, seeded and finely chopped
4 cloves garlic, finely chopped
3 stems lemon grass (white part only), finely chopped
2 teaspoons ground turmeric
10 candlenuts
2 large onions, chopped
1 cup (250 ml/8 fl oz) coconut milk
1.5 kg (3 lb) chicken, cut into 8 pieces
1/2 cup (125 ml/4 fl oz) coconut cream
2 tablespoons lime juice

1 Put the shrimp in a frying pan and dry-fry over low heat, shaking the pan regularly, for 3 minutes, or until the shrimp are dark orange and are giving off a strong aroma. Transfer to a mortar and pestle, and pound until finely ground. Alternatively, process in a food processor.

2 Place half the oil, the chilli, garlic, lemon grass, turmeric and candlenuts in a food processor, and process in short bursts until very finely chopped, regularly scraping down the side of the bowl.

3 Heat the remaining oil in a wok or frying pan, add the onion and 1/4 teaspoon salt, and cook, stirring regularly, over low heat for 8 minutes, or until golden.

4 Add the spice mixture and shrimp, and stir for 5 minutes. If the mixture begins to stick, add 2 tablespoons of the coconut milk. It is important to cook the mixture thoroughly to develop the flavours.

5 Add the chicken to the wok and cook, stirring, for 5 minutes, or until beginning to brown. Stir in the remaining coconut milk and 1 cup (250 ml/8 fl oz) water, and bring to the boil. Reduce the heat and simmer for 50 minutes, or until the chicken is cooked and the sauce has thickened slightly. Add the coconut cream and bring the mixture back to the boil, stirring constantly. Add the lime juice and serve immediately with rice.

LEMON AND ROSEMARY CHICKEN STEW

INGREDIENTS

8 large chicken drumsticks
60 g (2 oz) butter
2 cloves garlic, crushed
2 teaspoons finely grated lemon rind
2 tablespoons chopped fresh rosemary
1 tablespoon plain flour
1$^1/_2$ cups (375 ml/12 fl oz) chicken stock
2 tablespoons lemon juice

1 Using a sharp knife, make two deep cuts in the thickest part of each chicken drumstick.

2 Melt the butter in a large frying pan. Add the drumsticks and cook over medium heat for 2 minutes on each side, or until brown. Add the garlic, lemon rind and rosemary.

3 Blend the flour, stock and lemon juice until smooth. Add to the pan and bring to the boil. Reduce the heat and simmer, covered, for 25 minutes, or until the drumsticks are tender, stirring occasionally. Season, and serve immediately.

CHICKEN AND BACON GOUGERE

INGREDIENTS

60 g (2 oz) butter
1–2 cloves garlic, crushed
1 red onion, chopped
3 slices bacon, chopped
$^1/_4$ cup (30 g/1 oz) plain flour
1$^1/_2$ cups (375 ml/12 fl oz) milk
$^1/_2$ cup (125 ml/4 fl oz) cream
2 teaspoons wholegrain mustard
250 g (8 oz) cooked chicken, chopped
$^1/_2$ cup (30 g/1 oz) chopped fresh parsley

Choux pastry
60 g (2 oz) butter, chopped
$^1/_2$ cup (60 g/2 oz) plain flour
2 eggs, lightly beaten
$^1/_3$ cup (35 g/1$^1/_4$ oz) grated Parmesan

1 Melt the butter in a frying pan and cook the garlic, onion and bacon for 5–7 minutes, stirring occasionally, until cooked but not browned. Stir in the flour and cook for 1 minute. Gradually add the milk and stir until thickened. Simmer for 2 minutes, then add the cream and mustard. Remove from the heat and fold in the chopped chicken and parsley. Season with pepper.

2 To make the pastry, place the butter and $^1/_2$ cup (125 ml/4 fl oz) water in a pan. Stir until melted. Bring to the boil. Add the flour and beat for 2 minutes, or until the mixture leaves the side of the pan. Cool for 5 minutes. Gradually mix in the egg with an electric beater, until thick and glossy. Add the Parmesan.

3 Preheat the oven to hot 210°C (415°F/Gas 6–7). Grease a deep 23 cm (9 inch) ovenproof dish, pour in the filling and spoon heaped tablespoons of choux around the outside. Bake for 10 minutes, then reduce the oven to 180°C (350°F/Gas 4) and bake for 20 minutes, or until the choux is puffed and golden. Sprinkle with grated Parmesan.

CHICKEN AND LEEK PIE

INGREDIENTS

50 g (1¾ oz) butter
2 large leeks, washed and thinly sliced
4 spring onions, sliced
1 clove garlic, crushed
¼ cup (30 g/1 oz) plain flour
1½ cups (375 ml/12 fl oz) chicken stock
½ cup (125 ml/4 fl oz) cream
1 medium barbecued chicken, chopped
2 sheets puff pastry, thawed
¼ cup (60 ml/2 fl oz) milk

1 Preheat the oven to moderately hot 200°C (400°F/Gas 6). In a pan, melt the butter and add the leek, spring onion and garlic. Cook over low heat for 6 minutes, or until the leek is soft but not browned. Sprinkle in the flour and mix well. Pour in the stock gradually and cook, stirring well, until the mixture is thick and smooth.

2 Stir in the cream and the chicken. Put the mixture in a shallow 20 cm (8 inch) pie dish and set aside to cool.

3 Cut a circle out of one of the sheets of pastry to cover the top of the pie. Brush around the rim of the pie dish with a little milk. Put the pastry on top and seal around the edge firmly. Trim off any overhanging pastry and decorate the edge with the back of a fork. Cut the other sheet into 1 cm (½ inch) strips and roll each strip up loosely into a spiral. Arrange the spirals on top of the pie, starting from the middle and leaving a gap between each one. The spirals may not cover the whole surface of the pie. Make a few small holes between the spirals to let out any steam, and brush the top of the pie lightly with milk. Bake for 25–30 minutes, or until the top is brown and crispy. Make sure the spirals look well cooked and are not raw in the middle.

NOTE Make small pies by placing the mixture into 4 greased 1¼ cup (315 ml/10 fl oz) round ovenproof dishes. Cut the pastry into 4 rounds to fit. Bake for 15 minutes, or until crisp.

CHICKEN BALLOTINE

INGREDIENTS

1.6 kg (3¼ lb) chicken
2 red capsicums (peppers)
1 bunch (1 kg /2 lb) silverbeet
30 g (1 oz) butter
1 onion, finely chopped
1 clove garlic, crushed
1/2 cup (50 g/1¾ oz) grated Parmesan
1 cup (80 g/2 ¾ oz) fresh breadcrumbs
1 tablespoon chopped fresh oregano
200 g (6½ oz) ricotta

1 To bone the chicken, cut through the skin on the centre back with a sharp knife. Separate the flesh from the bone down one side to the breast, being careful not to pierce the skin. Follow along the bones closely with the knife, gradually easing the meat from the thigh, drumstick and wing. Cut through the thigh bone where it meets the drumstick and cut off the wing tip. Repeat on the other side, then lift the rib cage away, leaving the flesh in one piece and the drumsticks still attached to the flesh. Scrape all the meat from the drumsticks and wings, discarding the bones. Turn the wing and drumstick flesh inside the chicken and lay the chicken out flat, skin-side down. Refrigerate.

2 Preheat the oven to moderate 180°C (350°F/Gas 4). Quarter the capsicums and remove the membranes and seeds. Place skin-side up under a hot grill until the skin blisters and blackens. Place in a plastic bag and allow to cool, then peel off the skin.

3 Remove the stalks from the silverbeet and finely shred the leaves. Melt the butter in a large frying pan and add the onion and garlic. Cook over medium heat for about 5 minutes, or until soft. Add the silverbeet leaves and stir until wilted and all the moisture has evaporated. In a food processor, process the silverbeet and onion mixture with the Parmesan, breadcrumbs, oregano and half the ricotta. Season to taste with salt and cracked pepper.

4 Spread the silverbeet mixture over the chicken and lay the capsicum over the silverbeet. Form the remaining ricotta into a roll and place across the width of the chicken. Fold the sides of the chicken in over the filling so they overlap slightly, and tuck the ends in neatly. Secure with toothpicks and tie with string at 3 cm (1¼ inch) intervals.

5 Lightly grease a large piece of aluminium foil and place the chicken in the centre. Roll the chicken up securely in the foil, sealing the ends well. Place on a baking tray and bake for 1¼–1½ hours, or until the juices run clear when a skewer is inserted into the centre. Allow to cool, then refrigerate until cold before removing the foil, toothpicks and string. Cut into 1 cm (½ inch) slices to serve.

NOTE If you are not confident about boning a chicken, ask your butcher to do it for you. To be on the safe side, it might be worth ordering your boned chicken a day in advance.

CHICKEN BIRYANI

INGREDIENTS

Marinade

6 cardamom pods
3 onions, peeled
3 cloves garlic
5 cm (2 inch) piece ginger, sliced
8 cloves or 1/4 teaspoon ground cloves
1 teaspoon whole black peppercorns
1 teaspoon ground cumin
1 teaspoon ground cinnamon
1 1/2 tablespoons poppy seeds
1/4 teaspoon ground nutmeg
1 teaspoon salt
2 tablespoons lemon juice
250 g (8 oz) natural yoghurt

1 kg (2 lb) chicken pieces, cut into small pieces
1/3 cup (80 ml/2 3/4 fl oz) oil
3 bay leaves
2 whole cardamom pods, lightly crushed
3 onions, very thinly sliced
2 tablespoons raisins
1 1/2 cups (300 g/10 oz) long-grain white rice
1/4 cup (60 ml/2 fl oz) milk
1 teaspoon sugar
1 teaspoon saffron threads
plain yoghurt, to serve
1/4 cup (40 g/1 1/4 oz) toasted cashews, to serve

1 To make the marinade, crush the cardamom pods with the flat side of a large knife blade to release the seeds; discard the pods. Chop the onions, garlic and ginger in a food processor. Add the cardamom seeds, cloves, peppercorns, cumin, cinnamon, poppy seeds, nutmeg, salt and lemon juice, and process to a smooth paste. Stir in the yoghurt and set aside.

2 Prick the skin of the chicken with a fork (this helps the flavour infuse into the chicken) and place in a large mixing bowl. Add the marinade and mix well to thoroughly coat the chicken, then refrigerate for 3–4 hours, or overnight.

3 Heat about 3 tablespoons of the oil in a heavy-based frying pan over low heat. Add the bay leaves and whole cardamom pods and toss in the oil. Cook for 2 minutes, being careful not to burn. Remove the bay leaves and set aside; discard the cardamom pods. Add the onion and fry over low heat for 8 minutes, or until a rich golden colour. Remove with a slotted spoon, leaving the oil in the pan. Drain on paper towels. Quickly fry the raisins until plump; set aside with the onions. Reserve the remaining oil.

4 Place the chicken and the marinade in a heavy-based flameproof casserole dish and bring slowly to the boil. Cover and simmer for 15 minutes. Using tongs or a slotted spoon, remove the chicken, leaving as much marinade in the pan as possible. Reduce the marinade down, cooking it over low heat until only about 1 cup (250 ml/8 fl oz) remains. Return the chicken to the pan and toss to coat in the thickened sauce. Cover and set aside while preparing the rice.

5 Add the rice to a large pan of boiling water and cook for 7 minutes. (Do not cook the rice until tender.) Drain the rice into a colander. Meanwhile, heat the milk and sugar, pour into a bowl, add the saffron and soak for 5 minutes.

6 Preheat the oven to slow 150°C (300°F/Gas 2). Spoon the rice over the spice-coated chicken and pour over the saffron mixture. Streak the orange colour gently through the rice with a fork. Pour over the reserved flavoured oil, and scatter with bay leaves. Cover the dish tightly with foil and bake for 1 hour, or until the rice and chicken are tender. Serve topped with the onion, raisins, yoghurt and cashews.

CHICKEN WITH BEANS AND ASPARAGUS

INGREDIENTS

1 stem lemon grass, white part only, chopped
5 cm (2 inch) piece ginger, peeled and chopped
2–3 small red chillies, seeded and chopped
1 teaspoon grated kaffir lime (makrut) or lime rind
2–3 cloves garlic, chopped
1/2 teaspoon ground black pepper
2 tablespoons oil
375 g (12 oz) chicken breast fillets, cut into thin strips
250 g (8 oz) green beans, cut into short pieces
1 celery stick, cut into thick slices
185 g (6 oz) snow peas (mangetout), halved
200 g (6 1/2 oz) asparagus, cut into short pieces
270 ml (9 fl oz) can coconut cream
2 tablespoons sweet chilli sauce
20 small fresh basil leaves

1. Place the lemon grass, ginger, chilli, lime rind, garlic, pepper and oil in a food processor or blender, and process until the mixture forms a rough paste. Combine the paste and chicken strips in a glass or ceramic bowl, cover and refrigerate for at least 15 minutes.

2. Briefly blanch the beans, celery, snow peas and asparagus in a pan of boiling water. Drain and plunge into iced water. Drain again.

3. Heat the wok until very hot and stir-fry the chicken mixture in batches over high heat for 3–4 minutes, or until the chicken is cooked through. Stir constantly so the paste doesn't burn. Return all of the chicken to the wok with the vegetables, coconut cream, sweet chilli sauce, to taste, and basil leaves. Stir-fry until heated through. Serve with rice or noodles.

CHICKEN WITH OLIVES AND SUN-DRIED TOMATOES

INGREDIENTS

olive oil, for cooking
600 g (1 1/4 lb) chicken breast fillets, cut diagonally into thin slices
1 red onion, thinly sliced
3 cloves garlic, finely chopped
2 tablespoons white wine vinegar
1 teaspoon sambal oelek
1 tablespoon lemon juice
12 Kalamata olives, pitted and quartered lengthways
1/4 cup (40 g/1 1/4 oz) sun-dried (sun-blushed) tomatoes, cut into thin strips
1/4 cup (15 g/1/2 oz) finely chopped fresh parsley
1 tablespoon shredded fresh basil

1. Heat a wok until very hot, add 2 teaspoons of the oil and swirl it around to coat the side. Stir-fry the chicken slices in two batches until browned and cooked through, adding more oil in between each batch. Remove all the chicken from the wok and keep warm.

2. Reheat the wok, add 1 tablespoon of the oil and stir-fry the onion until it is soft and golden. Add the garlic and cook for 1 minute. Return the warm chicken to the wok. Add the vinegar, sambal oelek and lemon juice, and toss well.

3. Stir in the olive pieces, sun-dried tomato, parsley and basil, and season with salt and ground black pepper. Heat through thoroughly.

FAMILY CHICKEN PIE

INGREDIENTS

Pastry

2 cups (250 g/8 oz) self-raising flour
125 g (4 oz) butter, chopped
1 egg

Filling

1 barbecued chicken
30 g (1 oz) butter
1 onion, finely chopped
310 g (10 oz) can creamed corn
1¼ cups (315 ml/10 fl oz) cream

1 To make the pastry, process the flour and butter in a food processor for 15 seconds, or until the mixture is fine and crumbly. Add the egg and 2–3 tablespoons water and process for 30 seconds, or until the mixture just comes together. Turn onto a lightly floured surface and gather together into a smooth ball. Cover with plastic wrap and refrigerate for 20 minutes.

2 Meanwhile, to make the filling, remove the meat from the chicken carcass and shred finely. Heat the butter in a pan and cook the onion over medium heat for 3 minutes. Add the chicken, corn and cream. Bring to the boil, then reduce the heat and simmer for 10 minutes. Remove from the heat and allow to cool slightly.

3 Preheat the oven to moderate 180°C (350°F/Gas 4). Roll half the pastry between two sheets of plastic wrap to cover the base and side of a 23 cm (9 inch) pie dish. Spoon the chicken mixture into the pastry-lined dish.

4 Roll the remaining pastry to cover the top of the pie. Brush with milk. Press the edges together to seal. Trim the edges with a sharp knife. Roll the excess pastry into two long ropes and twist together. Brush the pie edge with a little milk and place the pastry rope around the rim. Bake for 45 minutes.

GINGER CHICKEN WITH MUSHROOMS AND WHEAT NOODLES

INGREDIENTS

4 dried Chinese mushrooms
2 teaspoons cornflour (cornstarch)
2 tablespoons soy sauce
2 tablespoons oyster sauce
1 tablespoon mirin or sweet sherry
200 g (6 1/2 oz) dried wheat noodles
1 teaspoon sesame oil
oil, for cooking
2–3 cloves garlic, crushed
8 cm (3 inch) piece fresh ginger, cut into matchsticks
375 g (12 oz) chicken breast fillets or tenderloins, cut into thin strips
1 red onion, cut into thin wedges
6 spring onions, cut into short lengths
185 g (6 oz) small field mushrooms, thickly sliced
1 cup (90 g/3 oz) bean sprouts
1/3 cup (20 g/3/4 oz) chopped fresh mint

1. Place the dried mushrooms in a small bowl and cover with hot water. Leave to soak for 10 minutes, or until softened. Drain and squeeze dry, then discard the hard centre stem and chop the mushrooms finely.

2. Combine the cornflour with 1/4 cup (60 ml/2 fl oz) water and mix to a fine paste. Add the soy sauce, oyster sauce and mirin.

3. Cook the noodles in a large pan of boiling salted water for 1–2 minutes, or according to the manufacturer's instructions. Drain and set aside.

4. Heat the wok until very hot, add the sesame oil and 1 tablespoon of the oil, and swirl it around to coat the side. Stir-fry the garlic, ginger and chicken strips in batches over high heat for 2–3 minutes, or until the chicken has cooked through. Remove from the wok and set aside.

5. Reheat the wok, add 1 tablespoon of the oil and stir-fry the red onion and spring onion for 1–2 minutes, or until softened. Add the dried and field mushrooms, then stir-fry the mixture for 1–2 minutes, or until tender. Remove from the wok and set aside.

6. Add the soy sauce mixture to the wok and stir for 1–2 minutes, or until the sauce is well heated and slightly thickened. Return the chicken and vegetables to the wok with the bean sprouts, noodles and chopped mint. Stir until the noodles are well coated with the sauce. Serve at once.

GOAN-STYLE CHICKEN AND SULTANAS AND ALMONDS

INGREDIENTS

2 teaspoons ground cumin
2 teaspoons ground coriander (cilantro)
1 teaspoon ground cinnamon
1/2 teaspoon cayenne pepper
1/2 teaspoon ground cardamom
oil, for cooking
1 large onion, cut into thin wedges
2 cloves garlic, finely chopped
500 g (1 lb) chicken breast fillets, cut into cubes
2 teaspoons finely grated orange rind
2 tablespoons orange juice
2 tablespoons sultanas
1 teaspoon soft brown sugar
1/4 cup (60 g/2 oz) thick plain yoghurt
1/3 cup (40 g/1 1/4 oz) slivered almonds, toasted

1. Dry-fry the spices in a wok over low heat for about 1 minute, or until fragrant, shaking the wok regularly.

2. Add 1 tablespoon oil to the wok and stir-fry the onion wedges and garlic over high heat for 3 minutes. Remove from the wok.

3. Reheat the wok, add 1 tablespoon of the oil and stir-fry the chicken in two batches until it is golden and just cooked. Return all the chicken to the wok with the onion mixture, orange rind, juice, sultanas and sugar. Cook for 1 minute, tossing until most of the juice evaporates.

4. Stir in the yoghurt and reheat gently, without boiling or the yoghurt will separate. Season well with salt and pepper. Serve garnished with the toasted almonds.

NOTE Yoghurt separates easily when heated due to its acid balance. Yoghurt also separates when shaken, whipped or stirred too much.

HONEY CHICKEN

INGREDIENTS

oil, for cooking
500 g (1 lb) chicken thigh fillets, cut into cubes
1 egg white, lightly beaten
1/3 cup (40 g/1 1/4 oz) cornflour (cornstarch)
2 onions, thinly sliced
1 green capsicum (pepper), cubed
2 carrots, cut into batons
100 g (3 1/2 oz) snow peas, sliced
1/4 cup (90 g/3 oz) honey
2 tablespoons toasted almonds

1 Heat a wok until very hot, add 1 1/2 tablespoons of the oil and swirl it around to coat the side. Dip half of the chicken into the egg white, then lightly dust with the cornflour. Stir-fry over high heat for 4–5 minutes, or until the chicken is golden brown and just cooked. Remove from the wok and drain on paper towels. Repeat with the remaining chicken, then remove all the chicken from the wok.

2 Reheat the wok, add 1 tablespoon of the oil and stir-fry the sliced onion over high heat for 3–4 minutes, or until slightly softened. Add the capsicum and carrot, and cook, tossing, for 3–4 minutes, or until tender. Stir in the snow peas and cook for 2 minutes.

3 Increase the heat, add the honey and toss the vegetables until well coated. Return the chicken to the wok and toss until it is heated through and is well coated in the honey. Remove from the heat and season well with salt and pepper. Serve immediately, sprinkled with the almonds.

ORANGE ROASTED CHICKENS

INGREDIENTS

2 x 800 g (1 lb 10 oz) chickens
100 g (3 1/2 oz) butter, softened
2 cloves garlic, crushed
1 tablespoon finely grated orange rind
1/2 cup (60 ml/2 fl oz) orange juice

1. Preheat the oven to hot 220°C (425°F/Gas 7). Using kitchen scissors, cut the chickens in half through the backbone and breastbone. Pat dry with paper towels and wipe the inside.

2. Combine the butter, garlic and orange rind and beat well. Gently loosen the skin of the chickens by sliding your fingers between the flesh and the skin. Push the orange butter under the skin as evenly as possible. Put the chickens onto a ceramic dish and pour on the orange juice. Cover and refrigerate for 3 hours, or preferably overnight.

3. Drain the chicken pieces well and arrange cut-side down on roasting racks inside two baking dishes. Pour 2 tablespoons of water into each baking dish.

4. Roast for 30–40 minutes, or until the chickens are golden brown. Cover with foil and allow to rest for 15 minutes. Cut into quarters to serve.

NOTE If you can, use freshly squeezed orange juice.

RICE WITH CHICKEN AND SEAFOOD

INGREDIENTS

500 g (1 lb) raw medium prawns (shrimp)
500 g (1 lb) mussels
200 g ($6\frac{1}{2}$ oz) calamari (squid) tubes
$\frac{1}{4}$ cup (60 ml/2 fl oz) oil
2 chorizo sausages, thickly sliced
500 g (1 lb) chicken pieces
300 g (10 oz) pork fillet, thickly sliced
4 cloves garlic, crushed
2 red onions, chopped
$\frac{1}{4}$ teaspoon saffron threads, soaked in hot water
$\frac{1}{4}$ teaspoon turmeric
4 large tomatoes, peeled, seeded and chopped
2 cups (440 g/14 oz) short-grain rice
1.25 litres hot chicken stock
125 g (4 oz) green beans, cut into 4 cm ($1\frac{1}{2}$ inch) lengths
1 red capsicum (peppers), cut into thin strips
1 cup (155 g/5 oz) fresh peas

1 Peel the prawns. Devein, leaving the tails intact. Scrub the mussels and remove the beards. Cut the calamari tubes into 5 mm ($\frac{1}{4}$ inch) thin slices. Heat 1 tablespoon of the oil in a large, heavy-based pan and add the chorizo. Cook over medium heat for 5 minutes, or until browned. Drain on paper towels. Add the chicken pieces and cook for 5 minutes, or until golden, turning once. Drain on paper towels.

2 Add the pork to the pan and cook for 3 minutes, or until browned, turning once. Drain on paper towels. Heat the remaining oil in the pan, add the garlic, onion, drained saffron and turmeric, and cook over medium heat for 3 minutes, or until the onion is soft. Add the tomato and cook for 3 minutes, or until soft.

3 Add the rice and stir for 5 minutes, or until the rice is translucent. Stir in the hot chicken stock, bring to the boil, cover and simmer for 10 minutes. Add the chicken, cover and cook for 20 minutes. Add the pork, prawns, mussels, calamari, chorizo and vegetables. Cover and cook for 10 minutes, or until the liquid has been absorbed.

STUFFED CHICKEN BREASTS

INGREDIENTS

1 tablespoon olive oil
1 onion, finely chopped
2 cloves garlic, crushed
100 g (3$^1/_2$ oz) ham, finely chopped
1 green capsicum (pepper), finely chopped
2 tablespoons finely chopped pitted black olives
$^1/_3$ cup (35 g/1$^1/_4$ oz) grated Parmesan
6 chicken breast fillets
plain flour, to coat
2 eggs, lightly beaten
1$^1/_2$ cups (150 g/5 oz) dry breadcrumbs
$^1/_4$ cup (60 ml/2 fl oz) olive oil

1 Heat the oil in a pan and add the onion, garlic, ham and capsicum. Cook, stirring, over medium heat for 5 minutes, or until the onion is soft. Remove and place in a heatproof bowl. Add the olives and Parmesan.

2 Cut a deep pocket in the side of each chicken fillet, cutting almost through to the other side.

3 Fill each fillet with the ham mixture and secure with toothpicks along the opening of the pocket. Coat each fillet with the flour, shaking off any excess. Dip into the beaten egg and then coat with the breadcrumbs. Heat the oil in a large pan and cook the fillets, in batches, over medium-high heat for 15–20 minutes, turning halfway through, until golden and cooked through. To serve, remove the toothpicks, then cut diagonally into thin slices.

VIETNAMESE CHICKEN AND NOODLE CASSEROLE

INGREDIENTS

1 stem lemon grass
4 kaffir lime (makrut) leaves
4 cups (1 litre/$5\frac{1}{3}$ pt) chicken stock
400 ml (13 fl oz) coconut cream
$\frac{1}{4}$ cup (30 g/1 oz) coconut milk powder
2 tablespoons peanut oil
400 g (13 oz) chicken breast fillets, cut into strips
12 raw king prawns (jumbo shrimp), peeled and deveined, tails intact
8 spring onions, sliced
2 teaspoons finely chopped fresh ginger
4 cloves garlic, finely chopped
2 small red chillies, seeded and finely chopped
500 g (1 lb) Hokkien noodles
1 teaspoon dried shrimp paste
2 tablespoons lime juice
1 cup (90 g/3 oz) bean sprouts
fresh mint leaves, to garnish
fresh coriander (cilantro) leaves, to garnish

1 Finely chop the white stem of the lemon grass. Remove the centre stem from the kaffir lime leaves, then finely shred the leaves.

2 Place the lemon grass and lime leaves in a large, heavy-based pan with the stock, coconut cream and coconut milk powder. Bring to the boil, stirring constantly to dissolve the coconut milk powder. Reduce the heat and simmer, covered, for 15 minutes.

3 Heat a wok over high heat and add the peanut oil. Add the chicken, prawns, spring onion, ginger, garlic and chilli. Stir-fry for 5–10 minutes, or until the chicken and prawns are cooked through.

4 Place the noodles in the simmering coconut cream, then add the chicken and prawn mixture from the wok. Add the shrimp paste and lime juice. Allow the noodles to heat through.

5 Divide the sprouts among warmed deep bowls and place the noodles, chicken and prawns on top. Ladle the sauce over, scatter with mint and coriander and serve at once.

BLACKENED CHICKEN WITH CRISPY TORTILLA

INGREDIENTS

4 vine-ripened tomatoes, cut into 1 cm (1/2 inch) slices
1 teaspoon caster sugar
1 red onion, sliced
150 ml olive oil
1 ripe avocado
60 g (1/4 cup) sour cream
100 ml (3 1/2 fl oz) milk
2 tablespoons lime juice
2 x 16 cm corn tortillas
1 teaspoon dried oregano
2 1/2 teaspoons ground cumin
1 1/4 teaspoons garlic salt
1/2 teaspoon cayenne pepper
4 small skinless chicken breast fillets (about 600 g/21 oz)
15 g (1/2 cup) coriander (cilantro) leaves

1 Place the tomato slices in a wide dish, sprinkle with sugar and season well. Layer the onion over the top and drizzle with 60 ml (1/4 cup) of oil. Chill for 20 minutes.

2 Blend the avocado, sour cream, milk, lime juice and 80 ml (1/3 cup) water in a food processor for 1 minute or until smooth. Season.

3 Cut each of the corn tortillas into eight 2 cm (1 inch) strips. Combine the oregano, cumin, garlic salt and cayenne pepper, and coat the chicken breasts in the spice mixture, pressing down firmly with your fingers. Heat 1 1/2 tablespoons of oil over medium– high heat in a large, non-stick frying pan until hot. Cook the chicken breasts for 4–5 minutes on each side, or until cooked through. Cool, then refrigerate. In the same pan add 60 ml (1/4 cup) of oil. Fry the tortilla strips until golden, turning once during cooking.

4 On each plate arrange the tomato and onion slices in a small circle. Slice each chicken breast on the diagonal into 2 cm (1 inch) pieces and arrange over the tomato. Spoon the dressing over the chicken and arrange four tortilla strips over the top. Sprinkle with coriander leaves and serve immediately.

CHARGRILLED CHICKEN WITH SPINACH AND RASPBERRIES

INGREDIENTS

60 ml ($^1/_4$ cup) raspberry vinegar
2 tablespoons lime juice
2 garlic cloves, crushed
2 tablespoons chopped oregano
1 teaspoon soft brown sugar
2 small red chillies, finely chopped
125 ml ($^1/_2$ cup) virgin olive oil
4 chicken breast fillets
1 teaspoon Dijon mustard
200 g baby English spinach leaves
250 g fresh raspberries

1 Mix 2 tablespoons of the raspberry vinegar, the lime juice, crushed garlic, 1 tablespoon of the oregano, the sugar, chilli and 60 ml ($^1/_4$ cup) of the oil in a large bowl. Immerse the chicken in the marinade, cover and refrigerate for 2 hours.

2 Preheat the oven to 180°C (350°F/Gas 4). Heat a chargrill pan and cook the chicken for 3 minutes on each side, then place on a baking tray and bake for a further 5 minutes, or until cooked through. Allow the chicken to rest for 5 minutes, then cut each breast into five strips on the diagonal.

3 To make the dressing, combine the remaining oil, vinegar and oregano with the mustard, $^1/_4$ teaspoon salt and freshly ground black pepper and mix well. Toss the spinach and raspberries with half of the dressing. Top with the chicken and drizzle with the remaining dressing.

MISO YAKITORI CHICKEN

INGREDIENTS

3 tablespoons yellow or red miso paste
2 tablespoons sugar
60 ml (1/4 cup) sake
2 tablespoons mirin
1 kg (2 lb 3 oz) chicken thighs, boned (skin on)
1 cucumber
2 spring onions, cut into 2 cm (3/4 inch) pieces

1. Soak 12 long wooden bamboo skewers in cold water for at least 10 minutes. Place the miso, sugar, sake and mirin in a small saucepan over medium heat and cook, stirring well, for 2 minutes, or until the sauce is smooth and the sugar has dissolved completely.

2. Cut the chicken into 2.5 cm (1 inch) cubes. Seed the cucumber and cut into 2 cm (3/4 inch) batons. Thread the chicken, cucumber and spring onion pieces alternately onto the skewers — you should have three pieces of chicken, three pieces of cucumber and three pieces of spring onion per skewer.

3. Cook on a chargrill plate over high heat, turning occasionally, for 10 minutes, or until the chicken is almost cooked. Brush with the miso sauce and continue cooking, then turn and brush the other side. Repeat this process once or twice until the chicken and vegetables are cooked. Serve immediately with rice and salad.

WILD RICE AND ROAST CHICKEN WITH ASIAN DRESSING

INGREDIENTS

190 g (1 cup) wild rice
200 g (1 cup) jasmine rice
1 Chinese barbecue roast chicken (see note)
15 g (1/4 cup) chopped mint
15 g (1/4 cup) chopped coriander (cilantro)
1 large Lebanese cucumber
6 spring onions
80 g (1/2 cup) roasted peanuts, roughly chopped
80 ml (1/3 cup) mirin
2 tablespoons Chinese rice wine
1 tablespoon soy sauce
1 tablespoon lime juice
2 tablespoons sweet chilli sauce, plus extra, to serve

1. Bring a large saucepan of water to the boil and add 1 teaspoon of salt and the wild rice. Cook for 30 minutes, add the jasmine rice and cook for a further 10 minutes, or until tender. Drain the rice, refresh under cold water and drain again.

2. Shred the chicken (the skin as well) into bite-sized pieces, place in a large bowl and add the mint and coriander. Cut the cucumber through the centre (do not peel) and slice thinly on the diagonal. Slice the spring onions on the diagonal. Add the cucumber, spring onion, rice and peanuts to the bowl with the chicken.

3. Mix together the mirin, rice wine, soy, lime juice and sweet chilli sauce in a small jug, pour over the salad and toss to combine. Pile the salad onto serving platters and serve with extra chilli sauce.

NOTE It is important to use a Chinese barbecued chicken, available from Chinese barbecue shops. The flavours of five-spice and soy used to cook it will add to the flavour of the dish.

CHICKEN BREASTS WITH MUSTARD CREAM SAUCE

INGREDIENTS

4 chicken breasts (about 200 g/7 oz each)
2 tablespoons oil
1 clove garlic, crushed
1/4 cup (60 ml) dry white wine
2 tablespoons wholegrain mustard
2 teaspoons chopped fresh thyme
300 ml (10 fl oz) cream
240 g (8 1/2 oz) green beans, topped and tailed
320 g (11 1/4 oz) baby yellow squash, halved

1 Pound each chicken breast between sheets of plastic wrap with a mallet or rolling pin until about 1 cm (1/2 inch) thick.

2 Heat the oil in a frying pan over high heat. Brown the chicken breasts for 4–5 minutes on each side, or until brown. Remove and cover with foil.

3 Add the garlic to the frying pan and cook for 1 minute over medium heat, then stir in the wine, mustard and thyme. Increase the heat to medium–high and pour in the cream. Simmer for about 5 minutes, or until the sauce has reduced and thickened slightly, then season to taste.

4 Meanwhile, bring a saucepan of lightly salted water to the boil, add the beans and squash and cook for 2–4 minutes, or until just tender. Season to taste. To serve, pour a little of the sauce over the chicken and serve with the vegetables on the side.

MEXICAN CHICKEN BAKE

INGREDIENTS

3/4 cup (165 g) short-grain rice
300 g (10 1/2 oz) can red kidney beans, drained and thoroughly rinsed
3 1/2 tablespoons chopped fresh coriander (cilantro) leaves
1 tablespoon oil
600 g skinless chicken thigh fillets, unrolled
2 x 200 g (7 oz) jars spicy taco sauce
2 cups (250 g/9 oz) grated Cheddar
1/2 cup (125 g/4 1/2 oz) sour cream

1. Preheat the oven to moderate 180°C (350°F/Gas 4). Lightly grease a deep (7 cm/3 inch) round (21 cm/8 inch) ceramic casserole dish. Bring a large saucepan of water to the boil, add the rice and cook for 10–12 minutes, stirring occasionally. Drain.

2. In the prepared dish, combine the beans and 1 1/2 tablespoons of the coriander, then add the rice and toss together. Lightly press the mixture so the beans are mixed into the rice and the mixture is flat.

3. Heat the oil in a large frying pan over medium–high heat. Sauté the chicken thighs for 3 minutes, then turn over. Add the spicy taco sauce, and cook for another 3 minutes.

4. To assemble, spread half the cheese over the rice. Arrange the thighs and sauce on top in a star shape, sprinkle with 1 1/2 tablespoons coriander, then sprinkle with cheese. Cover with foil.

5. Bake for 35–40 minutes, or until the mixture is bubbling and the cheese is melted and slightly browned—remove the foil for the last 5 minutes. Cut into four servings with a knife and scoop out carefully, keeping the layers intact. Serve sprinkled with the remaining coriander and a dollop of sour cream.

SPANISH SAFFRON CHICKEN AND RICE

INGREDIENTS

$^{1}/_{4}$ cup (60 ml/2 fl oz) olive oil
4 chicken thighs and 6 drumsticks
1 large red onion, finely chopped
1 large green capsicum (pepper), two thirds diced and one third julienned
3 teaspoons sweet paprika
400 g (14 oz) can diced tomatoes
1$^{1}/_{4}$ cups (275 g/9$^{3}/_{4}$ oz) paella or arborio rice
$^{1}/_{2}$ teaspoon ground saffron

1 Heat 2 tablespoons of the oil in a large deep frying pan over high heat. Season the chicken pieces well and brown in batches. Remove the chicken from the pan.

2 Reduce the pan to medium heat and add the remaining oil. Add the onion and the diced capsicum and cook gently for 5 minutes. Stir in the paprika and cook for 30 seconds. Add the tomato and simmer for 1–3 minutes, or until it thickens.

3 Stir 3$^{1}/_{2}$ cups (875 ml) boiling water into the pan, then add the rice and saffron. Return the chicken to the pan and stir to combine. Season, to taste. Bring to the boil, then cover, reduce the heat to medium–low and simmer for 20–30 minutes, or until all the liquid has been absorbed and the chicken is tender. Stir in the julienned capsicum, then allow to stand, covered, for 3–4 minutes before serving.

STUFFED CHICKEN BREAST WITH TOMATO, GOAT'S CHEESE AND ASPARAGUS

INGREDIENTS

4 large chicken breast fillets
100 g ($3\frac{1}{2}$ oz) semi-dried (sun-blushed) tomatoes
100 g ($3\frac{1}{2}$ oz) goat's cheese, sliced
200 g (7 oz) asparagus spears, trimmed, halved and blanched
50 g butter
$1\frac{1}{2}$ cups (375 ml/$12\frac{1}{2}$ fl oz) chicken stock
2 zucchini (courgettes), cut into 5 cm (2 inch) batons
1 cup (250 ml/$8\frac{1}{2}$ fl oz) cream
8 spring onions, thinly sliced

1. Pound each chicken breast between two sheets of plastic wrap with a mallet or rolling pin until 1 cm ($\frac{1}{2}$ inch) thick. Divide the tomato, goat's cheese and 155 g ($5\frac{1}{2}$ oz) of the asparagus pieces among the breasts. Roll up tightly lengthways, securing along the seam with toothpicks.

2. Heat the butter in a large frying pan over medium heat. Add the chicken, then brown on all sides. Pour in the stock, then reduce the heat to low. Cook, covered, for 10 minutes, or until the chicken is cooked through. Remove the chicken and keep warm.

3. Meanwhile, bring a saucepan of lightly salted water to the boil. Add the zucchini and remaining asparagus and cook for 2 minutes, or until just tender. Remove from the pan. Whisk the cream into the frying pan. Add the spring onion and simmer over medium–low heat for 4 minutes, or until reduced and thickened. To serve, cut each chicken roll in half on the diagonal and place on serving plates. Spoon on the sauce and serve with the greens.

TERIYAKI CHICKEN WITH GINGER CHIVE RICE

INGREDIENTS

4 small chicken breast fillets, skin on (about 170 g/6 oz each)
1/4 cup (60 ml/ 2 fl oz) Japanese soy sauce
2 tablespoons sake
1 1/2 tablespoons mirin
1 1/2 tablespoons soft brown sugar
3 teaspoons finely grated fresh ginger
1 1/2 cups (300 g/10 1/2 oz) long-grain rice
2 tablespoons finely chopped fresh chives
2 tablespoons oil

1. Pound each breast between sheets of plastic wrap with a mallet until 1 cm (1/2 inch) thick. Put the soy sauce, sake, mirin, sugar and 1 teaspoon ginger in a flat non-metallic dish and stir until the sugar has dissolved. Add the chicken and turn to coat. Cover and refrigerate for 1 hour, turning once halfway through.

2. Bring a large saucepan of water to the boil. Add the rice and cook for 12 minutes, stirring occasionally. Drain. Stir in the chives and remaining ginger, then cover until ready to serve.

3. Drain the chicken, reserving the marinade. Heat the oil in a deep frying pan and cook the chicken, skin-side-down over medium heat for 5 minutes, until the skin is crisp. Turn and cook for 4 minutes (not quite cooked).

4. Add the marinade and 1/4 cup (60 ml) water to the pan and scrape up any sediment. Bring to the boil over high heat, then add the chicken (skin-side-up) and juices. Cook for 5–6 minutes, until cooked through, turning once. (If the sauce is runny, remove the chicken and boil the sauce until syrupy.) Serve the chicken whole or sliced, drizzled with the sauce.

CHICKEN CASSEROLE WITH OLIVES AND TOMATO

INGREDIENTS

1 tablespoon olive oil
1 large onion, chopped
2 garlic cloves, crushed
8 pieces chicken, skin on
1 tablespoon tomato paste (purée)
375 ml (1 1/2 cups) white wine
a pinch of sugar
8 large ripe tomatoes, chopped
4 tablespoons parsley, chopped
180 g (6 1/2 oz) green beans, topped, tailed and halved
130 g (4 1/2 oz) olives

1 Heat the oil in a large flameproof casserole and fry the onion for a minute or two. Add the garlic and the chicken and fry for as long as it takes to brown the chicken all over.

2 Add the tomato paste and white wine, along with the sugar, and stir everything together. Add the tomato and any juices, the parsley and the beans and bring to the boil. Turn down the heat, season well and simmer for 40 minutes.

3 Add the olives and simmer for another 5 minutes. The sauce should be thick by now and the chicken fully cooked. Add more salt and pepper, if necessary. Serve with potatoes, pasta or rice.

AVGOLEMONO WITH CHICKEN

INGREDIENTS

1 carrot, chopped
1 large leek, chopped
2 bay leaves
2 chicken breast fillets
2 litres (8 cups) chicken stock
75 g ($^1/_3$ cup) short-grain rice
3 eggs, separated
80 ml ($^1/_3$ cup) lemon juice
2 tablespoons chopped parsley
40 g (1$^1/_2$ oz) butter, chopped

1 Place the carrot, leek, bay leaves, chicken fillets and stock in a large saucepan. Bring to the boil over high heat, then reduce the heat and simmer for 10–15 minutes or until the chicken is cooked. Strain into a clean saucepan and reserve the chicken.

2 Add the rice to the liquid, bring to the boil, then reduce the heat and simmer for 15 minutes, or until tender. Cut the chicken into 1 cm ($^1/_2$ inch) cubes.

3 Whisk the egg whites in a clean, dry bowl until firm peaks form. Beat in the yolks until light and creamy, whisk in the lemon juice, then 250 ml (1 cup) of the soup. Remove the soup from the heat and gradually whisk in the egg mixture. Add the chicken and stir over low heat for 2 minutes — do not boil or the egg will scramble. Serve at once with a sprinkle of parsley and dot of butter.

NOTE This soup will not stand well — make just before serving.

CHICKEN CASSEROLE WITH MUSTARD AND TARRAGON

INGREDIENTS

60 ml ($^{1}/_{4}$ cup) olive oil
1 kg (2 lb 4 oz) chicken thigh fillets, halved, then quartered
1 onion, finely chopped
1 leek, sliced
1 garlic clove, finely chopped
350 g (12 oz) button mushrooms, sliced
$^{1}/_{2}$ teaspoon dried tarragon
375 ml (1$^{1}/_{2}$ cups) chicken stock
185 ml ($^{3}/_{4}$ cup) cream
2 teaspoons lemon juice
2 teaspoons Dijon mustard

1 Preheat the oven to 180°C (350°F/Gas 4). Heat 1 tablespoon of the oil in a flameproof casserole dish over medium heat, and cook the chicken in two batches for 6–7 minutes each, or until golden. Remove from the dish.

2 Add the remaining oil to the casserole dish and cook the onion, leek and garlic over medium heat for 5 minutes, or until soft. Add the mushrooms and cook for 5–7 minutes, or until they are soft and browned and most of the liquid has evaporated. Add the tarragon, chicken stock, cream, lemon juice and mustard, bring to the boil and cook for 2 minutes. Return the chicken pieces to the dish and season well. Cover.

3 Place the casserole in the oven and cook for 1 hour, or until the sauce has reduced and thickened. Season with salt and pepper and serve with potatoes and a green salad.

PASTA WITH CHICKEN, MUSHROOM AND TARRAGON

INGREDIENTS

375 g (13 oz) fusilli, or other pasta shapes such as ruote, conchiglie or penne
2 tablespoons virgin olive oil
350 g (12 oz) chicken tenderloins, cut into 2 cm (3/4 inch) pieces
20 g (3/4 oz) butter
400 g (14 oz) Swiss brown or button mushrooms, sliced
2 garlic cloves, finely chopped
125 ml (1/2 cup) dry white wine
185 ml (3/4 cup) cream
1 teaspoon finely grated lemon zest
2 tablespoons lemon juice
1 tablespoon chopped tarragon
2 tablespoons chopped parsley
25 g (1/4 cup) grated Parmesan cheese, plus extra, to serve

1. Cook the pasta in a large saucepan of boiling salted water until al dente.
2. Meanwhile, heat 1 tablespoon of the oil in a large frying pan, add the chicken and cook over high heat for 3–4 minutes, or until lightly browned. Remove from the pan.
3. Heat the butter and the remaining oil, add the mushrooms and cook, stirring, over high heat for 3 minutes. Add the garlic and cook for a further 2 minutes.
4. Pour in the wine, then reduce the heat to low and simmer for 5 minutes, or until nearly evaporated. Add the cream and chicken and simmer for about 5 minutes, or until thickened.
5. Stir the lemon zest, lemon juice, tarragon, parsley and Parmesan into the sauce. Season with salt and pepper, then add the hot pasta, tossing until well combined. Serve with the extra Parmesan.

PASTA WITH ROAST CHICKEN, PINE NUTS AND LEMON

INGREDIENTS

1.3 kg (3 lb) chicken
1 garlic bulb, cloves separated and left unpeeled
60 ml (1/4 cup) olive oil
30 g (1 oz) butter, softened
1 tablespoon finely chopped thyme
125 ml (1/2 cup) lemon juice
500 g (1 lb 2 oz) bavette or spaghetti
2 tablespoons currants
1 teaspoon finely grated lemon zest
50 g (1/3 cup) pine nuts, toasted
15 g (1/2 cup) finely chopped flat-leaf (Italian) parsley

1. Preheat the oven to 200°C (400°F/Gas 6). Remove the neck from the inside of the chicken and place the neck in a roasting tin. Rinse the inside of the chicken with cold water and shake out any excess. Insert the garlic cloves into the cavity, then put the chicken in the roasting tin.

2. Combine the oil, butter, thyme and lemon juice, then rub over the chicken. Season the chicken. Roast for 1 hour, or until the skin is golden and the juices run clear when the thigh is pierced with a skewer. Transfer the chicken to a bowl to catch any juices while resting. Remove the garlic from the cavity, cool, then squeeze the garlic cloves out of their skins and finely chop.

3. Cook the pasta in a large pan of boiling salted water until al dente. Meanwhile, pour the juices from the roasting tin into a saucepan and discard the neck. Add the currants, zest and chopped garlic, then simmer over low heat. Remove all the meat from the chicken and shred into bite-size pieces. Add the resting juices to the pan.

4. Add the chicken meat, pine nuts, parsley and the sauce to the hot pasta and toss well. Season with salt and pepper and serve.

PASTA WITH ARTICHOKES AND CHARGRILLED CHICKEN

INGREDIENTS

1 tablespoon olive oil
3 chicken breast fillets
500 g (1 lb) pasta
8 slices prosciutto
280 g (9 oz) jar artichokes in oil, drained and quartered, oil reserved
150 g (5 oz) semi-dried tomatoes, thinly sliced
90 g (3 oz) baby rocket leaves
2–3 tablespoons balsamic vinegar

1 Lightly brush a chargrill or frying pan with the oil and heat over high heat. Cook the chicken for 6–8 minutes each side, or until cooked through. Thinly slice and set aside.

2 Cook the pasta in a large pan of rapidly boiling salted water until al dente. Drain and return to the pan to keep warm. Meanwhile, place the prosciutto on a grill tray and cook under a hot grill for 2 minutes each side, or until crisp. Cool slightly and break into pieces. Combine the pasta with the chicken, prosciutto, artichokes, tomato and rocket in a bowl and toss. Whisk together 1/4 cup (60 ml/2 fl oz) of the reserved artichoke oil and the balsamic vinegar and toss through the pasta mixture. Season and serve.

PENNE WITH SAUTEED CHICKEN, ASPARAGUS AND GOAT'S CHEESE

INGREDIENTS

500 g (1 lb) penne
350 g (11 oz) fresh asparagus spears
1 tablespoon olive oil
2 chicken breast fillets, cut into small cubes
1 tablespoon finely chopped fresh thyme
1 cup (250 ml/8 fl oz) chicken stock
1/3 cup (80 ml/2 3/4 fl oz) balsamic vinegar
150 g (5 oz) goats cheese, crumbled

1 Cook the pasta in a large pan of rapidly boiling salted water until al dente. Drain and return to the pan to keep warm.

2 Remove the woody ends from the asparagus, cut into short lengths and cook in a pan of boiling water for 3 minutes, or until just tender.

3 Heat the oil in a pan over high heat. Add the chicken and cook in batches, stirring occasionally, for 5 minutes, or until browned. Return all the chicken to the pan. Add the thyme and cook for 1 minute. Add the stock and vinegar and bring to the boil. Reduce the heat and simmer, stirring, for 3–4 minutes, or until the sauce has reduced slightly, then add the asparagus. Toss the pasta with the chicken in a serving bowl and sprinkle with the cheese. Season and serve.

SMOKED CHICKEN LINGUINE

INGREDIENTS

1 tablespoon olive oil
1 leek, thinly sliced
3 large cloves garlic, finely chopped
½ cup (125 ml/4 fl oz) dry white wine
300 g (10 oz) Swiss brown mushrooms, sliced
2 teaspoons chopped fresh thyme
300 ml (10 fl oz) thick cream
2 smoked chicken breast fillets, thinly sliced (see note)
350 g (11 oz) fresh linguine

1 Heat the oil in a saucepan. Add the leek and cook, stirring, over low heat for 3–4 minutes, or until soft. Add the garlic and cook for another minute. Pour in the wine and simmer for 2–3 minutes, or until the liquid has reduced by half.

2 Increase the heat to medium, add the mushrooms and thyme and cook for 5 minutes, or until any excess liquid has been absorbed, then add the cream and sliced chicken. Reduce the heat and simmer for 4–5 minutes, or until the sauce has slightly thickened.

3 Meanwhile, cook the pasta in a large pan of rapidly boiling salted water until al dente. Drain and divide among serving plates. Spoon on the sauce and serve.

NOTE Buy smoked chicken at the deli section of good supermarkets.

SPEEDY CHICKEN AND PASTA BAKE

INGREDIENTS

200 g (6½ oz) spiral pasta
425 g (14 oz) can cream of mushroom or broccoli soup
1 cup (250 g/8 oz) sour cream
1 teaspoon curry powder
1 barbecued chicken
250 g (8 oz) broccoli, cut into small pieces
1 cup (90 g/3 oz) fresh breadcrumbs
1½ cups (185 g/6 oz) grated Cheddar

1. Preheat the oven to moderate 180°C (350°F/Gas 4). Cook the pasta in a large pan of rapidly boiling salted water until al dente. Drain and return to the pan to keep warm.

2. Combine the soup, sour cream and curry powder and season to taste with freshly ground black pepper.

3. Remove the meat from the chicken and roughly chop. Combine the chicken with the pasta, broccoli and soup mixture. Spoon into four lightly greased 2-cup (500 ml/16 fl oz) ovenproof dishes and sprinkle with the combined breadcrumbs and cheese. Bake for 25–30 minutes, or until the cheese melts.

CHICKEN WITH LEMON AND CAPERS

INGREDIENTS

olive oil, for cooking
1 red onion, cut into thin wedges
25 g (3/4 oz) butter
800 g (1 lb 10 oz) chicken breast fillets, cut into bite-sized pieces
rind of 1 lemon, cut into thin strips
2 tablespoons baby capers, rinsed well and drained
1/3 cup (80 ml/2 3/4 fl oz) lemon juice
1/4 cup (15 g/1/2 oz) shredded basil

1. Heat the wok until very hot, add 2 teaspoons of the oil and swirl it around to coat the side. Add the red onion wedges and stir-fry until softened and golden. Remove from the wok and set aside.

2. Reheat the wok, add 2 teaspoons of the oil and half the butter, and stir-fry the chicken in two batches until it is browned, adding more oil and butter between batches. Return all the chicken to the wok with the onion.

3. Stir in the lemon rind, capers and lemon juice. Toss well and cook until warmed through. Add the shredded basil and season with salt and black pepper. Delicious served with creamy mashed potato.

CHICKEN WITH SNOW PEA SPROUTS

INGREDIENTS

2 tablespoons oil
1 onion, finely sliced
3 kaffir lime (makrut) leaves, shredded
3 chicken breast fillets, cubed
1 red capsicum (pepper), sliced
1/4 cup (60 ml/2 fl oz) lime juice
100 ml (3 1/2 fl oz) soy sauce
100 g (3 1/2 oz) snow pea (mangetout) sprouts
2 tablespoons chopped coriander leaves

1 Heat a wok over medium heat, add the oil and swirl to coat the base and side of the wok. Add the onion and kaffir lime leaves and stir-fry for 3–5 minutes, or until the onion begins to soften. Add the chicken and cook for a further 4 minutes. Add the capsicum and continue to cook for 2–3 minutes.

2 Stir in the lime juice and soy sauce and cook for 1–2 minutes, or until the sauce reduces slightly. Add the sprouts and coriander and cook until the sprouts have wilted slightly. Serve immediately.

CREAMY CHICKEN WITH TARRAGON

INGREDIENTS

5 chicken breast fillets, cut into slices
2 tablespoons oil
1 tablespoon chopped French tarragon
1 cup (250 ml/8 fl oz) cream
1 tablespoon lemon juice

1 Lightly season the slices of chicken with white pepper. Heat a wok over high heat, add the oil and swirl the wok to coat the base and side with oil. When the oil is hot, add the chicken, in batches, and cook over medium heat for 3–4 minutes, or until it is browned. Return all the chicken to the wok and stir through the tarragon.

2 Add the cream, bring to the boil and boil for 3 minutes, or until the sauce has thickened slightly. Add the lemon juice and season with salt and ground black pepper before serving.

NOTE If you are preparing this dish in advance, add the lemon juice after reheating to serve or the sauce will curdle. Avoid using Russian tarragon, which has a less subtle slightly aniseed flavour. Tarragon vinegar is a good substitute for lemon juice.

All our recipes are thoroughly tested in a specially developed test kitchen. Standard metric measuring cups and spoons are used in the development of our recipes. All cup and spoon measurements are level. We have used 60 g (2¼ oz/Grade 3) eggs in all recipes. Sizes of cans vary from manufacturer to manufacturer and between countries – use the can size closest to the one suggested in the recipe.

CONVERSION GUIDE

1 cup = 250 ml (9 fl oz)
1 teaspoon = 5 ml
1 Australian tablespoon = 20 ml (4 teaspoons)
1 UK/US tablespoon = 15 ml (3 teaspoons)

DRY MEASURES	LIQUID MEASURES	LINEAR MEASURES
30 g = 1 oz	30 ml = 1 fl oz	6 mm = ¼ inch
250 g = 9 oz	125 ml = 4 fl oz	1 cm = ½ inch
500 g = 1 lb 2 oz	250 ml = 9 fl oz	2.5 cm = 1 inch

CUP CONVERSIONS – DRY INGREDIENTS

1 cup almonds, slivered whole = 125 g (4½ oz)
1 cup cheese, lightly packed processed cheddar = 155 g (5½ oz)
1 cup wheat flour = 125 g (4½ oz)
1 cup wholemeal flour = 140 g (5 oz)
1 cup minced (ground) meat = 250 g (9 oz)
1 cup pasta shapes = 125 g (4½ oz)
1 cup raisins = 170 g (6 oz)
1 cup rice, short grain, raw = 200 g (7 oz)
1 cup sesame seeds = 160 g (6 oz)
1 cup split peas = 250 g (9 oz)

INTERNATIONAL GLOSSARY

capsicum	sweet bell pepper
chick pea	garbanzo bean
chilli	chile, chili pepper
cornflour	cornstarch
eggplant	aubergine
spring onion	scallion
zucchini	courgette
plain flour	all-purpose flour
prawns	shrimp
minced meat	ground meat

Where temperature ranges are indicated, the lower figure applies to gas ovens, the higher to electric ovens. This allows for the fact that the flame in gas ovens generates a drier heat, which effectively cooks food faster than the moister heat of an electric oven, even if the temperature setting is the same.

	°C	°F	GAS MARK
Very slow	120	250	½
Slow	150	300	2
Mod slow	160	325	3
Moderate	180	350	4
Mod hot	190(g)–210(e)	375–425	5
Hot	200(g)–240(e)	400–475	6
Very hot	230(g)–260(e)	450–525	8

Published in 2006 by Bay Books,
an imprint of Murdoch Books Pty Limited.

ISBN 1-92120-844-9
978-1-92120-844-7

Printed by Sing Cheong Printing Company Ltd.
Printed in China.